WATER RESOURCE PLANNING, DEVELOPMENT AND MANAGEMENT

US DRINKING WATER REGULATION

HISTORY AND POLITICS, 1914-2015

WATER RESOURCE PLANNING, DEVELOPMENT AND MANAGEMENT

Additional books in this series can be found on Nova's website under the Series tab.

Additional e-books in this series can be found on Nova's website under the eBooks tab.

WATER RESOURCE PLANNING, DEVELOPMENT AND MANAGEMENT

US DRINKING WATER REGULATION

HISTORY AND POLITICS, 1914-2015

MICHAEL ZARKIN

nova science publishers
New York

NOTICE TO THE READER

The Publisher has taken reasonable care in the preparation of this book, but makes no expressed or implied warranty of any kind and assumes no responsibility for any errors or omissions. No liability is assumed for incidental or consequential damages in connection with or arising out of information contained in this book. The Publisher shall not be liable for any special, consequential, or exemplary damages resulting, in whole or in part, from the readers' use of, or reliance upon, this material. Any parts of this book based on government reports are so indicated and copyright is claimed for those parts to the extent applicable to compilations of such works.

Independent verification should be sought for any data, advice or recommendations contained in this book. In addition, no responsibility is assumed by the publisher for any injury and/or damage to persons or property arising from any methods, products, instructions, ideas or otherwise contained in this publication.

This publication is designed to provide accurate and authoritative information with regard to the subject matter covered herein. It is sold with the clear understanding that the Publisher is not engaged in rendering legal or any other professional services. If legal or any other expert assistance is required, the services of a competent person should be sought. FROM A DECLARATION OF PARTICIPANTS JOINTLY ADOPTED BY A COMMITTEE OF THE AMERICAN BAR ASSOCIATION AND A COMMITTEE OF PUBLISHERS.

Additional color graphics may be available in the e-book version of this book.

Library of Congress Cataloging-in-Publication Data

ISBN: 978-1-53612-567-2

Published by Nova Science Publishers, Inc. † New York

Contents

PREFACE

What constitutes "safe" drinking water? For more than a century, the US government has attempted to answer this question by setting national standards for drinking water quality. In a federal system of governance, however, national standards only go so far. State and local governments have long considered it their prerogative to select water supplies and treatment technologies – decisions that largely determine whether or not national standards will ever be met. Tragedies like the drinking water crisis in Flint, MI remind us that there are definite limits to what federal power can achieve. Nevertheless, the quest to raise the quality of drinking water through national standards remains an important and underappreciated episode in the history of US public health policy.

In this book, Michael Zarkin traces the development of US drinking water standards, beginning with the earliest efforts by the US Public Health Service to craft national standards, and ending with the EPA's most recent efforts to implement the Safe Drinking Water Act. Along the way, Dr. Zarkin tells the story of the ideas, political battles, and scientific controversies that shaped our nation's drinking water regulations. In the end, Dr. Zarkin concludes that drinking water regulation is made through an unconventional style of politics not found in other areas of US environmental policy.

ACKNOWLEDGMENTS

This project would never have been completed without the generous assistance and support of so many others. The staff in the University of Utah library system provided much-needed assistance locating government documents. Westminster College, my employer, provided travel money that allowed me to workshop this research at several conferences. I also owe thanks to the many colleagues who reviewed my conference papers and journal submissions over the years. Their assistance helped to give the project direction, and elevated the quality of the final product.

This monograph incorporates text and ideas developed in journal articles published in the *Electronic Green Journal* (2015) and the *Public Organization Review* (2017). I offer a special thank you to the publishers of those journals, who granted permission to use material from those earlier studies.

Finally, I wish to thank my partner in life, Dr. Kim Zarkin, whose love, support, and encouragement inspire me to be a better person and a better professor.

LIST OF ABBREVIATIONS

AWWA	American Water Works Association
CCL	Contaminant Candidate List
EDF	Environmental Defense Fund
EPA	Environmental Protection Agency
GAO	General Accounting Office/Government Accountability Office
GAC	Granular Activated Carbon
HRL	Health Reference Level
MCL	Maximum Contaminant Level
MDEQ	Michigan Department of Environmental Quality
MHS	Marine Hospital Service
NAS	National Academy of Sciences
NDWAC	National Drinking Water Advisory Council
NPDWR	National Primary Drinking Water Regulation
NRC	National Research Council
NRDC	Natural Resources Defense Council
PCCL	Preliminary Contaminant Candidate List
PHS	Public Health Service
SAB	Science Advisory Board
SDWA	Safe Drinking Water Act

Chapter 1

INTRODUCTION

In April of 2014, the city of Flint, Michigan began drawing its drinking water from the Flint River. The decision to switch water supplies was made amid a severe fiscal crisis in Flint that led Michigan Governor Rick Snyder to appoint emergency managers to help get the city's affairs in order. The emergency managers determined that Flint could save $5 million if it stopped contracting with the City of Detroit for drinking water, and reactivated an aging water treatment plant that would use the Flint River as its raw water source (Graham, 2016).

Within weeks after the switch, however, the city faced a public health crisis that eventually caught the entire nation's attention. Flint residents quickly noted that the water coming out of their taps was murky and foul smelling. Testing revealed that the water contained unsafe levels of fecal coliform bacteria, prompting the water supplier to increase the amount of chlorine added to the water. The added chorine reacted with other compounds in the water to produce trihalomethanes—synthetic organic chemicals believed to be toxic in humans. Nevertheless, city officials insisted that the water issue was overblown, even as local residents continued to complain of unusual health effects (Graham, 2016).

In early 2015, officials from the US Environmental Protection Agency (EPA) and Michigan Department of Environmental Quality (MDEQ) became aware that there were also high levels of lead present in the water coming out of Flint residents' taps. Later in the year, a study by researchers from Virginia Tech University concluded that the lead contamination was caused by Flint River water, which was corroding the city's aging pipes. Local doctors pleaded with city officials to stop using the Flint River water after finding elevated levels of lead in the blood of children living in the area. After more than a year and a half of suffering the effects of a tainted water supply, Flint residents finally got the state to act: Governor Snyder reconnected Flint to the Detroit water supply, and appointed a task force to investigate the matter. In January of 2016, Governor Snyder declared a state of emergency in Flint, and President Obama followed suit shortly thereafter (Lin, Rutter, and Park, 2016).

When the governor's task force released its report in March of 2016, there was plenty of blame to go around. Errors in judgment were committed by agencies at all levels of government, but the task force placed the most blame at the feet of the MDEQ. The task force cited "cultural shortcomings" within the MDEQ that caused the agency to misapply or ignore federal guidelines and refuse to seek federal assistance in a timely manner. The task force further noted that the situation was made worse by the EPA's reluctance to interfere with state enforcement by insisting on proper lead treatment technologies (Flint Water Advisory Task Force, 2016).

In the end, these bureaucratic failings will cost the city and state hundreds of millions of dollars and impact the lives of Flint residents for decades to come—all in an effort to save $5 million.

Drinking Water: A Long Regulatory History

The Flint water crisis is a tragedy that could have been avoided through investments in infrastructure and more efficacious regulation. Fortunately, situations like the Flint crisis are rare in modern American history thanks to the elaborate regulatory system we have put in place. At the same time, however, Flint is an extreme reminder of exactly what can go wrong in a regulatory system like ours that requires federal, state, and local officials to cooperate and communicate effectively.

The US regulatory regime for drinking water protection was not the intellectual brainchild of any single great reformer, nor the result of any single carefully planned piece of legislation. Rather, the system we have today evolved in fits and starts over more than a century. Beginning in the 1880s, state and local governments gradually researched, developed, and implemented drinking water treatment technologies to protect the public health in urban population centers. In 1914, the federal government entered the fray when it issued the first national drinking water regulation: a quality standard aimed at preventing epidemics by ensuring that the drinking water served on boats and trains was as pure as possible. Over the next fifty years, an informal division of labor gradually emerged between the federal and state governments: the federal government developed a fairly comprehensive, but largely voluntary set of drinking water quality standards, and the states attempted to meet those standards by adopting the most feasible water treatment technologies. In the 1970s, Congress passed the Safe Drinking Water Act (SDWA), and amended the law several more times over the next two decades. These statutory enactments, however, were largely an extension of the regime that was already in place. Congress added new national mandates, but the basic arrangement remains the same: the federal government sets national quality standards, but states expect a certain amount of autonomy in choosing how to meet those standards.

Perhaps the historical development of US drinking water regulation can be written off as a run-of-the-mill example of how the US federal system works. After all, federal-state collaboration is the norm today in so many areas of public policy. This study, however, is premised on the belief that US drinking water regulation has a unique policy history that raises a range of important, and heretofore unanswered, scholarly questions. Why, for instance, did the federal government choose to set a national drinking water quality standard in 1914—a time when national regulatory standards for public health and environmental protection were virtually unknown? Furthermore, what caused the federal and state governments to develop a cooperative, intergovernmental system of drinking water regulation decades before there was a federal statutory mandate to do so? Why did Congress find it necessary to pass the SDWA in 1974, decades after federal and state bureaucrats started developing a regulatory system on their own? Finally, why did Congress succeed in dramatically reforming the SDWA in the 1996, when efforts to reform other federal environmental and public health statutes were stalled?

To date, these questions have gone unanswered because scholars of US environmental policy have paid relatively little attention to drinking water regulation. Most scholarship by political scientists and legal scholars has focused only on isolated episodes in the history of US drinking water policy (Tarlock, 1997; Zarkin, 2015, 2017). Additionally, brief chapters outlining drinking water policy history may occasionally be found in volumes written for scientists and practitioners (Pontius, 2003). Up until now, however, no single volume has ever provided the comprehensive, historical examination of US federal drinking water policy needed to answer the questions mentioned in the preceding paragraph.

This study has two overarching purposes; first, to chronicle the historical development of US federal drinking water policy over the last century, and second, to explain the political forces that drove policy change during that same timespan. The historical narrative presented in the following four chapters adequately fulfills the first purpose. Fulfilling the second purpose, however, requires the introduction of political science theory into the discussion. Among scholars of US environmental policy, three main theoretical perspectives have gained currency—the society-centered, state-centered, and policy discourse perspectives. These three perspectives are outlined in the following section and will be employed further at the conclusion of this study.

Three Perspectives on Policymaking

In their basic form, the society-centered, state-centered, and policy discourse perspectives are by no means exclusive to the study of environmental policymaking. As was previously noted, however, these theories have been the basis for some of the most developed studies of national environmental policy. Although this study proceeds on the premise that the three perspectives are theoretically distinct, it is worth noting upfront that they share some basic commonalities. First, all three theories are capable of explaining changes in national environmental policy over an extended period of time. Second, all three perspectives decline to focus exclusively on any single "stage" of the policy process (agenda-setting, formulation, implementation), instead preferring to focus on the interactions of various political actors within a policy community. Nevertheless the perspectives differ in that each would prefer to view one particular set of explanatory factors—interests, the state, or ideas—as the primary catalyst for policy change.

In the remainder of this section, the three perspectives are elaborated. Admittedly, space limitations prevent an exhaustive treatment of each perspective. For this reason, the discussions that follow will be limited in two ways. First, the discussions will focus on the studies within each perspective that are most relevant to national environmental policymaking. Second, no effort is made to critique each perspective. Rather, the main goal is to isolate core tenets from each perspective that can be used to analyze the case study that follows in subsequent chapters.

The Society-Centered Perspective

Most major studies of US national environmental policy stress the importance of societal forces (interest groups, public opinion) in shaping policy outcomes. According to these authors, dramatic events (Birkland, 1997), shifts in public opinion (Jones, 1974), and changing economic conditions (Vogel, 1989) help to bring environmental issues onto the public agenda, but the actual formulation of new policies only proceeds when interest groups take the political stage. This is the case for two main reasons. First, many environmental policies impose concentrated costs while delivering diffuse benefits. As such, those in society who must absorb the costs (i.e., business interests) have strong incentives to organize against environmental policies, while those who receive the benefits (i.e., the mass public) constitute a larger, less organized group (Olson, 1965; Wilson, 1974). Furthermore, environmental policies tend to be scientifically complex, making it difficult for the average member of the public to actively participate in the policy formulation process (Gormley, 1986; Rosenbaum, 2014). As a result, environmental policy disputes frequently play out as negotiations between environmental advocacy groups and more conservative economic interests.

When society-centered theorists disagree, it is generally over the balance of power among contending interests (Kamieniecki, 2006; Kraft and Kamieniecki, 2007). Therefore, while some studies have concluded that the environmental movement exerted substantial influence in the late 1960s and early 1970s (Jones, 1974; Bosso, 1987; Harris and Milkis, 1996), others believe that America's individualistic political values (Cahn, 1995) and powerful economic elite (Gonzalez, 2001) prevented environmental groups from obtaining more far-reaching policy reforms. Yet others argue that the relative strength of environmental groups and business interests fluctuates over time in response to changing economic conditions, societal events, new issue frames, and the mobilization of new groups (Vogel, 1989).

To sum up, empirical tests of the society-centered perspective should confirm the proposition that interest groups, supported by other forces external to the state apparatus, not only set the policy agenda but also determine the contents of public policy. If the aforementioned proposition is confirmed, Society-centered perspectives may further seek to explore the balance of power among contending groups.

The State-Centered Perspective

Although society-centered studies acknowledge that state decision-makers and institutional arrangements shape policy debates, they generally believe that "the state" does not function independently of societal forces. State-centered studies, by contrast, identify two main ways in which state actors and institutions exert autonomous influence over the course of public policy. First, state theorists argue that existing institutional arrangements (laws, bureaucratic implementation regimes) shape the political participation opportunities available to societal groups, and limit the range of policy ideas and proposals that policymakers may legitimately choose to pursue in the future (Weir and Skocpol, 1985; Weir, 1992). Of particular concern to state theorists are the "policy legacies" that limit the prospects of fundamental policy reform. In essence, once a particular approach to solving a public problem becomes embedded in laws and bureaucratic enforcement practices, elected officials and interest groups generally do not want to take the political risks associated with completely dismantling existing policy frameworks. As a result, policy reformers end up taking the more incremental step of layering new legal requirements and enforcement programs on top of existing ones. For instance, Klyza (1996) concluded that even the most concerted efforts by powerful environmental groups in the 1960s and 1970s did not result in a fundamental restructuring of US resource management policies. Major policy changes were enacted during that time, but they mainly added to, rather than supplanted, existing policy goals and implementation strategies.

Second, state theorists find occasional evidence of "state autonomy"—situations where state officials "formulate and pursue goals that are not simply reflective of the demands or interests of social groups, classes, or society" (Skocpol, 1985, p. 9). Autonomous state action, however, depends on whether or not the state possesses the leadership capacity and internal expertise to formulate and implement effective policies. For instance, both Klyza (1992) and Carpenter (2001) concluded that the US Forest Service was able to autonomously determine the direction of forestry policy in the early 1900s because it earned a reputation as a trusted and highly competent bureaucratic entity under the leadership of Chief Forester Gifford Pinchot. Milazzo (2006) likewise demonstrated that policy entrepreneurs in Congress like Edmund Muskee harnessed the expertise possessed by bureaucrats and congressional staff to formulate and pass water pollution control legislation years before interest groups were actively engaged in the issue.

In short, the state-centered perspective is confirmed when there is evidence that policy legacies shape the course of future policy development, and state actors pursue autonomous goals and interests in the formulation of public policy.

The Policy Discourse Perspective

Society-centered and state-centered perspectives both treat policy ideas as important political tools that can be used by interest groups and state actors to frame issues and provide intellectual guidance in the formulation of policy. For discourse theorists, however, ideas and language have the potential to be truly transformative. Discourse theorists contend that many, if not most, environmental policy disputes play out between contending "discourses" or "languages," both of which amount to different ways of talking about policy problems (Dryzek, 2005; Williams and Matheny, 1995). Because environmental problems are complex, citizens, policymakers, scientists, and interest groups either cannot make sense of all relevant knowledge, or choose to interpret it in vastly different ways depending on their interests and worldviews. Policymaking participants attempt to overcome these political divisions by developing a common discourse—a way of framing a policy problem that enables opposing groups to see their interests in new ways and find common ground (Litfin, 1994; Hajer, 1995).

Discursive constructions of this kind are produced and reproduced in articles, reports, testimony, and other interactions between members of a policy community. According to Hajer (1995, pp. 58-61), successful discursive concepts lead to the formation of "discourse coalitions," or loose collections of policymaking participants who utilize a common set of concepts to express their points of view. Discourses become truly influential, however, when "discourse hegemony" is achieved. Discourse hegemony means that a particular discourse has become institutionalized within the state through laws and routines, and policymaking participants must utilize the established discourse in order to be perceived as credible.

For the purposes of this study, confirmation of the discourse perspective requires evidence that a particular set of discursive concepts was not only used in policy debates but also influenced the content of public policy.

RESEARCH DESIGN

The decision to both chronicle and explain the history of US drinking policy presupposes that drinking water politics is in some way distinct from other areas of environmental policy. Indeed, if the political dynamics of drinking water policy were the same as those discovered by scholars who have studied air pollution (Jones, 1974; Bryner, 1995; Harris and Milkis, 1996), surface water quality (Milazzo, 2006; Hoornbeek, 2011), pesticides (Bosso, 1987), and public lands (Klyza, 1996), then explanation would seem unnecessary. The politics surrounding drinking water policy, however, is somewhat different than in these other policy domains. Unlike other areas of environmental policy, drinking water regulation does not directly pit environmentalists against for-profit "polluting" industries. The drinking water industry consists mainly of a large number of small public and non-profit water systems that make pollution control their main business activity. As such, drinking water policy is made in a political environment where there is ostensibly less political disagreement between environmentalists and the regulated industry (Zarkin, 2015, 2017). For this reason, drinking water regulation appears to be a somewhat unique case deserving of explanation in its own right.

The 100-year timespan covered in this book provides an adequate data set for explaining policy change within a single area of policy. As Lijphart (1971) notes, longer timespans allows the comparison of multiple "data points" over a period of time, increasing the overall reliability of a single case study. Additionally, longer timespans more easily allow for the comparison of competing theoretical explanations. As Swift and Brady (1994, p. 83) point out, longer timespans provide fertile ground for testing competing explanations because they encompass the "range of variables, conditions, and outcomes … more likely to be representative of the true nature of political phenomena." Therefore, by examining drinking water regulation over a broader timespan, it is possible to gain a more complete picture of the forces that drive policy change.

PLAN OF THE STUDY

Chapters two through five chronicle the historical development of US drinking water policy. Chapter two chronicles drinking water policy development prior to 1970, starting with the earliest state-level efforts to improve drinking water quality, and ending with federal efforts to develop a comprehensive set of drinking water quality standards in the 1960s. Chapter three examines the formulation and implementation of the federal SDWA in the 1970s and early 1980s. Chapter four provides a discussion of the efforts made by Congress and the EPA to find a workable strategy for reforming the SDWA in the 1980s and early 1990s. Chapter five details the passage of the SDWA Amendments of 1996, and the EPA's subsequent implementation activities over the following two decades. The concluding chapter revisits the theoretical perspectives presented in this chapter to determine which one best explains the historical development of US drinking water regulation.

Chapter 2

BEGINNINGS OF FEDERAL INVOLVEMENT

The need for national drinking water standards grew out of the challenges posed by infectious disease. At the turn of the twentieth century, state governments invested in drinking water treatment technologies to combat the public health crises created by waterborne diseases like typhoid fever. To further prevent the spread of typhoid fever, The US Public Health Service (PHS) imposed the first national drinking water standard in 1914: a limitation on the amount of coliform bacteria allowed in water served to interstate travelers. Over the next five decades, the PHS standards were revised and expanded to include more than twenty bacterial and chemical contaminants. Although the states were not legally bound to enforce PHS standards against municipal water systems, the standards were eventually accepted as industry-wide benchmarks for the attainment of quality drinking water. In essence, by 1962 the US had a reasonably comprehensive, de facto set of national drinking water standards.

The PHS drinking water standards were remarkable for two overarching reasons. First, in 1914 the federal government generally did not have the capacity to actively participate in science-based regulatory standard setting. As Skowronek (1982) noted in his classic study of American state building, prior to the 1930s the national government lacked the bureaucratic apparatus to implement solutions to most complex social problems. The field of public health, however, proved to be a partial exception to this pattern. Concerned about the importation of disease from foreign nations, Congress set up a public health bureaucracy quite early in the nation's history and gradually funded the research capabilities and scientific expertise needed to solve public health problems. Therefore, when the time came for the federal government to address drinking water contamination, the issue found a natural home in an existing, reputable bureaucratic apparatus.

Second, the PHS went about the business of creating drinking water standards in spite of the fact that it possessed little legal authority to do so. The PHS claimed the authority to create the 1914 bacterial standard under a vague, general provision of the Interstate Quarantine Act relating to control of contagious and infectious disease. These same provisions, however, provided scant justification for chemical pollution standards. Nevertheless, the PHS was able to stretch the limits of its legal authority because the drinking water standards largely remained voluntary, and because they were created through a collaborative, advisory committee process that included a range of policy stakeholders.

This chapter chronicles the development of the nation's earliest federal drinking water standards. In order to do this, however, it is first necessary to consider the societal circumstances that preceded federal action.

Drinking Water and Infectious Disease

Western societies have been aware of the potential impact of drinking water on human health for thousands of years. Long ago, Hippocrates admonished citizens of Greece to carefully consider the sources from which their drinking water originated (McDermott, 1973). Ancient Romans, for their part, built elaborate aqueducts to channel potable water into populous urban areas (Baker, 1948). These early efforts to manage drinking water quality, however, were based on practical experience rather than use of the scientific method. It wasn't until well into the nineteenth century that a scientific link between drinking water and infectious disease was established.

Perhaps the first modern inquiry into the linkage between drinking water and public health was the epidemiological work of Dr. John Snow and his associates in the 1850s. Snow sought to ascertain the cause of a cholera outbreak that rapidly led to hundreds of deaths in a fairly isolated part of London. Through interviews with local residents, Snow figured out that virtually all of the afflicted residents drew their drinking water from a single pump located on Broad Street near St. James Parish. Once the pump was disabled, new instances of cholera declined within a week's time. Further investigation indicated that the local water supply was contaminated by sewage leaching into the well water from homes along Broad Street (Cosgrove, 1909).

Snow's investigations strongly suggested a link between human waste and the spread of infectious diseases such as cholera. The investigations of Louis Pasteur and other pioneers of bacteriology, however, more fully demonstrated cause and effect. Pasteur posited that bacteria caused many infectious diseases that afflicted humans and animals. Building on Pasteur's work, Robert Koch later isolated the specific microorganisms that were the likely causes of cholera, diphtheria, and typhoid fever in humans. The invention of vaccinations and antiseptics provided hope that infectious diseases might someday be cured. These diseases, however, spread rapidly when bacteria were transferred from human feces to drinking water, forcing scientists and public health officials to search for treatment technologies that would prevent exposure from occurring in the first place (Sedgwick, 1902).

THE PROMISE OF TREATMENT

Several notable outbreaks of cholera occurred in the US during the late 19th century, but typhoid fever proved to be a much more significant public health problem. In 1885, a typhoid epidemic in Plymouth, PA caused roughly 1000 infections and led to more than 100 deaths. In Chicago, nearly 3500 people died from typhoid fever between 1890 and 1892. In Pittsburgh, typhoid remained a public health crisis throughout the 1890s (Whipple, 1908). Overall, the rate of typhoid infection remained high in major US cities well into the first decade of the twentieth century (Ellms, 1917).

By the 1880s, the connection between typhoid fever and drinking water was well known. Urbanization led to a concentration of sewage waste that was neither being treated nor properly isolated from the drinking water supply. As a result, typhoid-causing bacteria easily spread to drinking water through human waste, quickly leading to outbreaks of epidemic proportions (Ellms, 1917). In the 1880s, the most immediate way to stop or prevent typhoid outbreaks was to draw water from a more distant source that was not yet infected with the disease. However, as more and more local water sources became infected through improper sewage disposal techniques, switching water sources became expensive and impractical. As a result, state and local public health officials sought out treatment technologies that would render polluted water sources potable (Tarr, Curley, McMichael, and Yosie, 1984).

Scientists and public health officials quickly latched on to sand filtration as the most affordable and effective treatment technology. Research in England and Germany demonstrated that when sewage was filtered through soil, biological processes ensued that prevented the groundwater below from becoming polluted. These results encouraged American scientists to conduct experiments that mimicked natural processes by allowing water to trickle through man-made filters filled with soil or fine sand particles (Sedgwick, 1902).

The technology of “slow sand filtration,” as it was called, was by no means novel in the 1880s. The first known sand filter was constructed in 1829 for the Chelsea Water Company in London, and over the next forty years sand filtration methods came into common usage in cities throughout Europe (Hazen, 1900). At that time, the health benefits of filtration were not known, and the technology was mainly used to clarify water through the elimination of sediment. Seeking ways to improve the appearance and quality of American water, the St. Louis, MO Board of Water

Commissioners sent engineer James P. Kirkwood to Europe to investigate filtration practices in the 1860s. Kirkwood examined filtration practices in seventeen European cities, and his subsequent report encouraged local governments in the US to invest in filtration technology (Kirkwood, 1869). In the mid-1870s Poughkeepsie, NY became the first US city to construct a drinking water filter and, by 1880, some 30,000 people throughout the US received filtered water (Johnson, 1914). Therefore, by the time researchers became interested in the health benefits of filtration, the technology was available and relatively easy to replicate.

The first major public health studies of filtration commenced in Massachusetts, when the state legislature appropriated a large sum of money for the state board of health to set up an experiment station in the city of Lawrence. Civil engineer Hiram F. Mills managed the Lawrence Experiment Station, but the experiments were conducted under the direction of MIT Professor William T. Sedgwick (Hazen, 1914). Sedgwick was a key figure in the early history of bacteriology and a proponent of "sanitary science," which revolved around the belief that science and human ingenuity could be harnessed to design the technologies needed to prevent infectious diseases. Sedgwick later noted in his influential textbook on sanitary science that "if [disease] be disturbance of a physical mechanism … their control … becomes easy and in a broad sense a problem of engineering" (Sedgwick, 1902, pp. viii-ix). To this end, Sedgwick taught one of the earliest courses on sanitary engineering at MIT, and the Lawrence Experiment Station became a training ground for some of the most prominent sanitary engineers of the day (Tarr, Curley, McMichael, and Yosie, 1984).

At the Lawrence Experiment Station, Sedgwick and his associates examined the impact of different types of sand and gravel on the purity of finished water. Researchers ultimately concluded that fine grained sand did the best job of purifying water, so long as periods of water filtration were followed by the injection of air to keep beneficial microorganisms alive in the sand (Sedgwick, 1902). When an epidemic of typhoid fever hit the city of Lawrence in the early 1890s, Sedgwick's sand filtration methods were put into practice, and new typhoid cases decreased by more than eighty percent within five years (Hazen, 1900). Following the success at reducing typhoid deaths in Lawrence, many other municipalities turned to intermittent, slow sand filtration as the preferred treatment technology.

Slow sand filtration, however, had its liabilities. As the name implies filtration is slow, making it difficult to purify adequate amounts of water for a densely populated area. The advent of rapid sand filtration helped to mitigate this problem by combining larger grains of sand with chemical coagulants that helped remove impurities from the water before it was filtered. The result was a faster filtration process, which yielded water roughly equal in purity to that produced through slow sand filtration techniques (Johnson, 1914). Neither filtration technique, however, fully eliminated the transmission of typhoid fever through drinking water. By the first decade of the twentieth century, sanitary engineers believed that supplementing filtration with chlorination and other disinfection techniques might completely eliminate the threat of disease (Jennings, 1923). In 1908, Jersey City, NJ became the first city to add chlorine compounds to its water, but other municipalities soon followed suit with great success. Minneapolis virtually eliminated the threat of typhoid fever by adding chlorine to drinking water after filtration was complete (McLaughlin, 1914). By the second decade of the twentieth century, sanitary engineers and state public health officials were optimistic that rapid deployment of filtration and disinfection technologies would end waterborne bacterial infections.

From State to Federal Involvement

Up until 1914, the problem of ensuring safe drinking water was entirely a state and local matter. By 1905, thirty-six states had passed legislation protecting drinking water, and most states followed Massachusetts in creating public health boards with authority over drinking water (Tarr, Curley, McMichael, and Yosie, 1984). Furthermore, state and local governments took the lead in experimenting with the treatment technologies that promised to improve drinking water safety. Nevertheless, there were limits to what state and local governments alone could do. Some states and localities were quicker than others to adopt cutting edge treatment technologies, making the quality of drinking water variable throughout the US until well into the twentieth century. This meant that if the water served on trains and ships was procured from a local source of inferior quality, passengers could be infected and disease could continue to spread throughout the country. For this reason, some form of action by the federal government was necessary to prevent the spread of disease in interstate commerce.

Prior to the 1930s the federal government was ill equipped to extend its authority into many areas of economic and social life because it lacked an established bureaucratic enforcement apparatus. The area of public health protection, however, was a partial exception to this pattern. The PHS began operation in the 1790s as the Marine Hospital Service (MHS)—a network of hospitals aimed at providing medical care to seamen serving the US on the high seas. From quite early on, a central mission of the MHS was to protect the public health through "quarantine," or the isolation of infected persons from the rest of the population. Though the power to quarantine was never explicitly mentioned in the US Constitution, it was recognized as a legitimate responsibility of national governments at the time of the founding (Williams, 1951). Following the ratification of the Constitution, Congress more or less assumed that quarantine responsibilities were inherent in its power to regulate interstate and foreign commerce. As a result, Congress passed numerous Quarantine Acts during the 19th Century to protect the American public against infectious disease (US Treasury Department, 1903).

During the 1880s, the MHS gradually added public health research capabilities to its arsenal of disease prevention tools. The MHS set up its Hygienic Laboratory in 1887 as little more than a one-room laboratory, but by the time that Congress reorganized the MHS as the PHS in 1902, the Laboratory was on its way to becoming the most important scientific research facility operated by the federal government (Harden, 1986). This transformation was due to the leadership of Milton J. Rosenau, who served as Director of the Hygienic Laboratory from 1899 to 1909. During those years, Dr. Rosenau got the Laboratory's facilities upgraded and its budget increased. More importantly, however, Rosenau and his staff conducted original research that led the PHS to be more proactive in the protection of public health.

In particular, research undertaken at the Hygienic Laboratory eventually led the PHS to conclude that an interstate standard for drinking water quality was necessary. Between 1906 and 1909, the Hygienic Laboratory undertook a major investigation to determine the causes of typhoid fever in the District of Columbia. Data collected as a part of the study indicated that typhoid was at least partially imported into DC through interstate travel (US Treasury Department, 1909). The DC study was recognized as an important milestone in the development of epidemiological research and, when Rosenau stepped down as director of the Lab in 1909, he was replaced by the study's director, John F. Anderson.

It was during Anderson's tenure at the Hygienic Laboratory that PHS undertook the task of writing a standard to govern the quality of drinking water served in interstate commerce. The PHS claimed that it was granted the legal authority to write such a standard by the Quarantine Act of 1893. The law vaguely granted the PHS the authority to promulgate national regulations "to prevent the introduction of contagious or infectious diseases into the United States, and into one state or territory or the District of Columbia from another state or territory or the District of Columbia" (US Treasury Department, 1903, p. 4). Though the language was imprecise and did not specifically relate to drinking water, the law did appear to support the development of such a standard for the limited purposes of controlling the bacterial quality of water served on interstate boats and trains.

The development of the drinking water standard was undertaken through an advisory committee consisting of PHS officials, academic researchers, and representatives from state boards of health. John Anderson chaired the committee, which also included William Sedgwick and Milton Rosenau, who by that time was serving as a Professor of Preventative Medicine and Hygiene at Harvard University. George Whipple, one of

Sedgwick's disciples, was also included in the deliberations. The committee spent more than a year and a half considering the available evidence and arriving at an appropriate standard. The final product was a quantitative limitation on the amount of coliform bacteria discovered when drinking water samples were subjected to specified laboratory testing methods (US Public Health Service [PHS], 1914).

In recommending the standard, the advisory committee was careful not to deviate from the best available scientific information and treatment practices. Coliform bacteria was not specifically the cause of typhoid fever, but became the basis of the water standard because it was easier to test for, and served as an indicator that other types of bacteria were likely present in the water supply. The committee also viewed the standard as attainable using readily available treatment technologies. As the committee noted in its progress report, "the art of water purification has … progressed to the point where there is no longer any excuse whatsoever for using bacteriologically polluted water" (PHS, 1914, p. 2964). Finally, the committee considered, but ultimately rejected the prospect of setting standards for chemicals and other potential water pollutants. The committee brushed aside chemical pollution as of "minor sanitary importance," instead characterizing bacterial pollution as not only "the most dangerous of all impurities," but also easier to deal with because "a vast deal of careful study" made it "possible to define with reasonable precision what constitutes a bacterially safe water supply" (pp. 2963-64). In essence, the advisory committee was unwilling to go out on a limb given the limits of both scientific knowledge and the legal mandate granted to the PHS.

From the outset, the 1914 drinking water standard was limited in key respects. First, the standard was not mandatory, and only applied to providers who wanted to sell water to rail and ship lines. Second, the advisory committee's report provided few details as to how the standard was to be enforced. Trains and boats travelling in interstate commerce were required to buy their water from local water providers who met the bacteriological standard. The PHS, however, did not have the resources to test water supplies or demand compliance from state and local governments. In essence, it was up to state boards of health to verify that water providers were regularly testing water and making that information available to their customers. In theory, the standard might well have simply been ignored until some major public health crisis forced Congress and the state governments to appropriate the money needed for enforcement activities. Remarkably, though, this did not happen. By the early 1920s, the standard was being followed by municipalities that did not sell water in interstate commerce, and was used by some courts as a legal standard for judging water purity (PHS, 1922). In essence, the water industry and state and local governments were informally adopting the 1914 water standard as a general industry standard for water quality.

The revelation that the 1914 drinking water standard was taking on industry-wide significance led the PHS to revise and update the rule in the 1920s. In 1922, another advisory committee was assembled, this time consisting not only of academics and public health officials, but also of representatives from six federal agencies and numerous scientific and professional associations (PHS, 1922). The committee's recommendations emphasized "safety" as the hallmark of quality drinking water. The committee stated that "safe" drinking water existed when "the danger, if any, is so small that it can not be discovered by any available means of observation" (PHS, 1925, p. 694). Although this language sounds somewhat precautionary in its orientation, the PHS stuck with its previous pattern of setting standards that were attainable given available treatment technologies and scientific testing procedures. As the PHS further indicated, the ultimate determinant of safety was reduction in disease, and the "best class" of water providers would serve as a reference point for setting acceptable standards. Water providers were first and foremost encouraged to use the highest quality water sources, and to protect sources from pollution whenever possible. Water providers were also encouraged to use treatment technologies, but to think of them as a precautionary measure rather than the sole basis for providing safe water. Ultimately, however, failure to meet the standard only meant that a water provider could not sell its water in interstate commerce.

THE RESPONSE TO CHEMICAL POLLUTANTS

The focus on protecting source waters became particularly important as the PHS began shifting its focus from bacterial to chemical pollutants in the 1940s. The decision to regulate a broad array of chemical pollutants was somewhat problematic given the limited legal authority possessed by the PHS. Under the Quarantine Act, the agency's legal mandate remained limited to the prevention of "contagious" and "infectious" diseases in interstate commerce. The chronic health effects associated with chemicals, though potentially quite serious, stretched this legal mandate to its virtual breaking point. An individual might be exposed to chemical contaminants by drinking water in interstate commerce, but a disease like cancer could certainly not be considered contagious or infectious. Furthermore, it was doubtful that the amount of exposure to chemicals experienced during the time the average individual spent on a train or boat would be enough to cause chronic disease. It was clear by this time that PHS's legal mandate to protect the public health in interstate commerce was incidental—the agency's true purpose was to create a set of standards that could serve as benchmarks for drinking water quality throughout the US. To this end, the PHS promulgated and revised standards for more than twenty chemical contaminants over the next several decades.

The first step occurred in 1925, when the PHS announced with little fanfare that excess levels of lead, copper, and zinc in drinking water would be considered grounds for rejection of a water source for use in interstate commerce. Quantitative limits for other chemical and mineral substances such as iron, sulfate, and magnesium were also set, though failure to meet these standards was not considered grounds for rejecting a water supply (PHS, 1925). At that time, the main purpose behind limiting chemical contaminants was the preservation of proper color, taste, and odor in the water. Although the PHS did indicate that "intestinal distress" might result from some of these chemicals, and vaguely used the phrase "toxic salts" to refer to others, discussion of the potential health effects of chemical contaminants was generally lacking from agency documents. Bacterial contamination remained the main focus of PHS drinking water activities for many years.

When the PHS once again revised the drinking water standards in 1942, it placed greater emphasis on chemical contaminants. Arsenic, selenium, and fluoride were added to the list of chemicals that triggered rejection of a water supply (PHS, 1943a). Once again, the PHS provided little explanation for its decision, noting only its firm conviction "that poisonous or otherwise harmful elements or salts in significant quantities … should not be allowed in water for drinking or culinary purposes" (PHS, 1943b, p. 104). Furthermore, the PHS provided no indication that treatment technologies were readily available for the removal of these chemical elements from water sources. Nevertheless, these standards were not viewed by water providers as an insurmountable obstacle: in 1946, the American Water Works Association—the leading industry trade association—for the first time issued a statement formally endorsing the PHS drinking water standards as industry-wide quality standards (PHS, 1946).

Although the PHS gave few specific reasons for the increased attention it placed on chemical contaminants, the decision was by no means obscure. Beginning in the 1920s, the federal government gradually became more concerned about the potential link between chemicals and cancer. During the 1920s and 1930s, scientists in the Division of Industrial Hygiene at the PHS pioneered methods of industrial epidemiology that enabled them to better understand the health consequences of exposure to chemicals in the workplace. In 1936, Congress created the National Cancer Institute (NCI) within the PHS and appropriated money to aid in cancer research. During the 1950s, Dr. Wilhelm Hueper, director of the NCI's section on environmental cancer, promoted the use of animal testing to further understand the link between chemical exposure and cancer in humans (Sellers, 1997). Concerns over the health effects of chemical pesticides and food additives led Congress in 1958 to pass the Delaney Clause, which amended the Food, Drug, and Cosmetics Act of 1938 to forbid the Secretary of Health, Education, and Welfare from approving food additives that were shown to cause cancer (Merrill, 1997).

As evidence grew that chemical exposure correlated with instances of cancer and other diseases, PHS officials became interested in developing quantitative drinking water standards specifically aimed at mitigating these negative health effects. Compared to bacterial standards, however, setting meaningful quantitative limits for chemical contaminants seemed like a herculean task. Bacterial infections were visible in human blood and waste, and the diseases they induced manifested themselves within days. By contrast, cancer and other chronic health consequences associated with chemical exposure took years or even decades to materialize, making the cause and effect link between chemicals and disease hard to pin down. As a result, scientists found it difficult to agree on "safe" levels of exposure to most chemicals (Kotin and Hueper, 1955).

These difficulties were further compounded by the fact that there was very little research examining the link between cancer and drinking water contaminants. Nor, for that matter, was there extensive research documenting the occurrence of a broad range of industrial chemicals in drinking water supplies throughout the US. Neither of these limitations, however, seemed to deter PHS officials from seeking to devise drinking water quality standards for chemical contaminants. In an article published in a leading trade journal, PHS scientists Herbert Stockington and Richard Woodward (1958) posited that it was possible to derive quantitative drinking water standards from existing toxicological research. Essentially, they argued that quantitative drinking water standards could be derived from three existing sources of knowledge: research on human tolerance of pesticide residue, toxicity data from nutrition studies, and existing quantitative standards set for chemical air pollutants in an occupational setting. In spite of their optimism, however, Stockington and Woodward warned of the difficulties associated with attempting to derive drinking water quality standards from these data, noting that the factors that needed to be considered were "far more numerous and complex than those involved in setting threshold limits for air tolerances or pesticides residue" (p. 524).

The approach suggested by Stockington and Woodward was quite a departure from the conservative approach taken by the PHS when the first bacterial standard was created in 1914. At that time, the PHS was unwilling to take steps that were not fully supported by existing scientific knowledge and treatment technologies. With chemical contaminants, however, Stockington and Woodward suggested an approach that better resembles the modern "precautionary principle," which encourages governments to be proactive about mitigating potential public health threats even if the available scientific evidence is incomplete (Weiss, 2007). This new precautionary approach seemed to garner the support of other participants in the drinking water standard setting process as well. Abel Wolman (1960), a prominent figure in the sanitary engineering field and a member of three of the four major PHS drinking water standard setting committees, also made a strong case for a precautionary approach to chemical contaminants:

> Obviously, the best criteria for detecting a dangerous public water supply are the doctor's certificate showing the man is dead and the epidemiological evidence showing that the water he drank killed him. The health official does not wait for such criteria, specific and quantitative as they are. They cannot be applied promptly enough, and they do not provide a wide area of protection. The health official chooses to widen this area immensely by moving to far less specific criteria with broad empiric relationships to disease (p. 1346).

This new precautionary mindset helped provide the impetus for revision of the drinking water standards once again in the early 1960s.

The drinking water standards promulgated in 1962 were by far the most comprehensive and carefully justified set of standards ever produced by the PHS. As a matter of law, the 1962 standards were no more far reaching than previous ones in that they only officially covered water served in interstate commerce. This fact, however, was lost amid the elaborate sixty-page report issued by the PHS that reads as though the standards were meant to have broad national applicability. The report detailed quantitative limitations for nearly two dozen chemicals, most of which were accompanied by narratives describing the sources of the pollutant, known health effects, and the scientific rationale behind the final standard. Each contaminant narrative was also accompanied by an extensive bibliography listing the most prominent scientific studies relating to each chemical (PHS, 1962).

As with each of the preceding sets of drinking water standards, the PHS relied mainly on an advisory committee consisting of federal and state public health officials, academic researchers, and representatives of the affected industries. A group of PHS scientists served as the committee's technical subcommittee, helping to assemble research for the committee to consider. The committee carefully considered available data concerning the health effects of various contaminants as well as their likely occurrence in drinking water (Hopkins, 1961). Admittedly, available data concerning health effects were incomplete, and the range of chemicals actually occurring in drinking water was unknown. For these reasons, the PHS committed itself to a process of continuous learning and more frequent revisions of standards. As the PHS stated upon announcing its intent to create the new standards, it would do its best to "derive … values from the best information" and adjust "upward or downward as new and better data became available" (PHS, 1961, p. 936).

CONCLUSION

By 1962, a de facto pattern had emerged in US drinking water policy. The national government set quantitative drinking water quality standards, but it was left up to state and local governments to implement them. Although this arrangement was never formally written into law, it became an institutionalized reality that was accepted by officials at all levels of government. General acceptance of this arrangement occurred because the PHS drinking water standards were largely voluntary, and because they were promulgated through an advisory committee process that made it possible to achieve consensus.

Over time, political scientists would eventually come to refer to this kind of federal-state collaboration as "cooperative federalism," and highlight the tensions that it created as the two levels of government vie for control over policy (Anton, 1989). Those tensions, however, would not manifest themselves until the 1970s, when changes in federal law started to make the cooperative federal relationship more coercive. The Safe Drinking Water Act of 1974, discussed in the next chapter, exemplified this new brand of federal coercion that would eventually lead to conflict within the drinking water policy community.

Chapter 3

FORMULATING AND IMPLEMENTING THE SDWA

With the passage of the Safe Drinking Water Act (SDWA) in late 1974, Congress replaced the sixty-year history of voluntary PHS standards with a regime of mandatory national standards that applied to the vast majority of public water systems in the US. The SDWA was part of a wave of federal legislation passed in the 1960s and 1970s that came to be known as the "new social regulation." Most of these new regulatory statutes were passed because a new generation of citizen activists mobilized and demanded legislation to curb the environmental, health, and safety threats created by corporate capitalism. The SDWA, however, was a partial exception to this pattern—it was not the pet project of a movement for social reform, but rather the brainchild of government officials who sought to enhance the federal role in drinking water protection at a time when the nation's attention was focused on environmental issues.

By 1974, federal authority over drinking water now rested with the newly created EPA. The SDWA provided the EPA with a mandate to create a broader, more stringent set of standards, but did not completely discard the policymaking patterns institutionalized during the PHS years. The federal government would continue to devise quantitative pollution limits, while states and localities would take the lead in matters related to enforcement, treatment technology, and the selection of drinking water sources. One new wrinkle introduced under the SDWA was the possibility that the EPA might occasionally mandate treatment technologies for public water systems when quantitative pollution limits were not practical. Mandating treatment technologies, however, was a risky proposition that threatened to upset established political expectations and create new lines of conflict within the drinking water policy community. Indeed, when the EPA chose to mandate a treatment technology for organic contaminants, the relationship between the federal government and water providers became adversarial for the first time in history.

THE NEW SOCIAL REGULATION

The phrase "new social regulation" was meant to contrast federal regulatory statutes passed in the 1960s and 1970s with the earlier forms of economic regulation that predominated during the first half of the twentieth century. Whereas economic regulation was focused on establishing economic security for infrastructure industries like transportation, communications, and banking, social regulation attempted to mitigate negative byproducts of the capitalist system such as environmental degradation, workplace and product safety issues, and social injustice (Eisner, 2000).

Although the phrase "new social regulation" suggests a drastic departure from past regulatory practices, these laws were actually a layering of old and new policy ideas and institutional arrangements (Harris and Milkis, 1996). On the one hand, the new social regulation reflected the Progressive Era faith in bureaucratic rationality. Early twentieth century Progressives believed that non-partisan bureaucratic experts were best positioned to solve industrial-era social and economic problems. To this end, Congress created regulatory agencies such as the Interstate Commerce Commission, Federal Trade Commission, and Federal Communications Commission, and endowed them with broad authority to regulate selected industries (McCraw, 1985; Horwitz, 1989). The new social regulation continued this pattern; Congress created a litany of new agencies, and authorized them to use science-based regulation to solve society's most pressing social problems (Bryner, 1987).

On the other hand, the new social regulation was also shaped by new ideas and institutional innovations inspired by the "New Left" political philosophy of the 1960s. New Left thought rejected both Soviet Communism and traditional American Liberalism, instead seeking to build a better democracy based on citizen participation and direct action. At the heart of the New Left agenda was a desire to construct new institutional arrangements that allowed citizens to hold government and corporate bureaucrats accountable (Harris and Milkis, 1996). Inspired by New Left ideas, American university students went on to found citizen groups in the 1960s and 1970s that raised public awareness and lobbied for passage of the new social regulatory statutes (Vogel, 1989).

New social regulatory statutes were a peculiar melding of these two seemingly contradictory governing philosophies. Congress created new bureaucratic entities like the EPA and gave them substantial authority to regulate. Congress also emphasized the importance of science in the regulatory process. Agencies like the EPA and the Occupational Safety and Health Administration were given science advisory boards to provide external review of agency scientific analysis (Jasanoff, 1990). At the same time, however, Congress wrote regulatory statutes that made these new agencies responsible to both legislators and citizen groups. To ensure that Congressional intent was followed, agencies were given very specific directions and timetables for writing regulations. To ensure accountability to citizens, some of the laws contained provisions allowing citizens to file suit against the agencies when proper enforcement actions were not being taken (Eisner, 2000).

In many respects the national pollution control laws passed during the 1970s were among the crown jewels of the new social regulation. Pollution control laws, like most all of the new social regulatory laws, were a direct outgrowth of public demand. The need to control industrial pollution became a topic of public discussion in the early 1960s when Rachel Carson highlighted the harms associated with chemical pesticides in her book

Silent Spring. Over the next several years, numerous focusing events helped keep environmental issues on the agenda. In 1967, the British oil tanker Torrey Canyon burst open, creating an environmental catastrophe that was broadcast around the world. Two years later, when the Cuyahoga River burst into flames near Cleveland, OH, Americans were forced to confront the poor state of the country's waterways (Andrews, 1999). Young activists formed new environmental groups such as the Natural Resources Defense Council (NRDC) and the Environmental Defense Fund (EDF) to fight for pollution control legislation (Bosso, 2005). Meanwhile, public opinion was shifting solidly in favor of increased governmental action. In April of 1970, the first Earth Day helped cement public opinion, when people around the country came together to discuss, learn about, and take action on environmental issues (Rosenbaum, 2014).

Sensing the broad national consensus that had developed around environmental issues, the federal government snapped into action. In 1970, President Nixon issued an executive order creating the EPA and transferring to it all authority possessed by the PHS and other federal agencies over matters relating to pollution control (Moe, 1989). The purpose behind the EPA's creation was clear: the US needed a single, unified pollution control agency with the resources to take on industrial polluters. Over the next decade, Congress enhanced the EPA's mission by passing a range of new pollution control statutes for the agency to enforce.

At the heart of the EPA's regulatory authority are the "command and control" regulatory statutes: the Clean Air Act Amendments (1970), the Federal Water Pollution Control Act Amendments (1972), the Safe Drinking Water Act (1974), and the Resource Conservation and Recovery Act (1976). These laws all follow a similar structure. The EPA first issues "commands" or national standards aimed at limiting the amount of pollution allowed into the air, land, and water. In the case of the Clean Air Act, for instance, the EPA is required to set National Ambient Air Quality Standards that mandate quantitative limits for specific chemical pollutants. Similarly, under the Safe Drinking Water Act, the EPA is required to set Maximum Contaminant Levels for drinking water pollutants. Once standards are issued, it becomes necessary to exert "control" over polluters through permitting, inspections, and other enforcement actions. All four of these laws are mainly enforced through a cooperative federalism arrangement, under which state governments become the primary enforcement agent once they demonstrate they are prepared to take on these responsibilities (Rosenbaum, 2014).

DRINKING WATER EMERGES ON THE AGENDA

When placed within their broader social context, it is tempting to view the pollution control laws passed between 1970 and 1976 as part of a single "policy regime" that coalesced around common goals and political forces (Fiorino, 2001; Jochim and May, 2010). This is true to a certain extent, but it is also overly simplistic. Although the early 1970s was a period of punctuated change in pollution control policy (Baumgartner and Jones, 2009), the decision to have separate laws to protect air, surface water, and drinking water was no accident. By the early 1970s, each of these issues was an established federal policy domain, with its own unique political and legal history.

Prior to the 1970s, Congress had been actively considering solutions to air and surface water pollution for decades. In both cases, federal legislation was already in place, and the laws passed in the early 1970s merely built on this legacy (Milazzo, 2006). In the case of drinking water, the history was somewhat different. Prior to 1971, Congress never held a hearing on drinking water issues or actively considered legislation to regulate drinking water quality. And why would they? By the 1940s a consensus had developed around voluntary national standards, formulated with input from the regulated industry and implemented by the states. Unlike other areas of environmental protection, there was simply no national political controversy surrounding drinking water, and no demand for federal legislation. When Congress finally took up the issue of drinking water quality, however, decades of inaction proved to be a liability. Drinking water standards were a scientifically complex matter, and Congress faced a steep learning curve if it intended to legislate in this area.

Therefore, it is not surprising that the regulatory regime ultimately enacted through the Safe Drinking Water Act of 1974 emulated and built upon the existing standards and governing arrangements created by the PHS and the states.

Drinking water issues found their way onto the congressional agenda in the early 1970s as a result of a series of studies that questioned the general quality of drinking water delivered by US public water systems. In 1970, the PHS released the results of its *Community Water Supply Study,* in which it surveyed 969 public water supply systems serving eight metropolitan areas and the state of Vermont. The study compared the water quality in these systems against PHS standards for the quality of drinking water served in interstate commerce. Using federal standards as a metric, the study found that 41% of the drinking water systems in the survey did not meet the PHS prescribed limitations for one or more contaminants. In addition, 90% of the systems in the survey did not meet PHS criteria for bacterial monitoring (PHS, 1970).

Within a few years, additional studies raised concerns about the link between cancer and chemical contaminants in drinking water. An EPA water survey found Mississippi River water near the city of New Orleans to be in exceptionally poor condition. As a result of years of industrial waste disposal, high levels of cyanide, mercury, arsenic, lead, and other dangerous chemicals and heavy metals were now present in the river water. In addition, the EPA found that finished drinking water taken from the river contained multiple organic contaminants that were believed to be carcinogenic. Taken together, these findings caused the EPA to conclude that the current condition of the Mississippi River posed a "potential threat to the health of 1.5 million people who consume this water" (EPA, 1972, p. 7). EPA's fears were seemingly confirmed two years later, when a study conducted on behalf of the EDF showed a positive correlation between rates of cancer and consumption of river water among Louisiana residents (Page, Harris, and Epstein, 1976).

Additional research conducted around that same time indicated that the potentially dangerous organic compounds found in Mississippi River water were a byproduct of the very chlorination processes designed to protect the public health. Chlorine interacted with hydrocarbons in drinking water to produce chloroform and other trihalomethanes that were suspected of causing cancer in humans. However, while it was believed that treatment technologies could somewhat reduce the instance of trihalomethanes in drinking water, safe levels of exposure to these chemicals were not known at the time. Furthermore, it was hypothesized that interactions between trihalomethanes and other chemicals found in drinking water might enhance or diminish cancer risks in ways that were not fully understood (Marx, 1974).

Taken together, these studies demonstrated that there was still much for scientists to learn about the link between cancer and chemical occurrence in drinking water. These limitations, however, did not deter members of Congress from introducing drinking water legislation. In fact, for Representatives Paul Rogers (D-FL) and Howard Robison (R-NY), two of the principal sponsors of what became the SDWA, the scientific uncertainty presented a welcome challenge. Speaking about his proposed legislation on the floor of the US House of Representatives, Robison conceded that "a broad program of research and development" was needed before US drinking water supplies could be effectively managed, but he expressed confidence that these obstacles could be overcome because "we have priceless expertise and technology in this nation which can give us answers to these questions" (*Safe Drinking Water*, 1971, pp. 63-64). Guided by this spirit of optimism, legislative consideration of drinking water legislation commenced.

LEGISLATIVE ACTION

Legislative deliberations commenced in 1971 when Reps. Rogers and Robison each introduced similar pieces of legislation requiring the EPA to set mandatory national drinking water standards for virtually all public water systems in the United States. Both bills also anticipated a stronger federal role in the enforcement of the standards. In May of that year, a lengthy set of hearings commenced in which legislators heard the perspectives of EPA officials, state regulators, and water industry trade associations. These early hearings revealed that key stakeholders in the drinking water policy community were not nearly as enthusiastic about the short-term prospect of federal legislation as the bills' sponsors.

EPA Administrator William Ruckleshaus was among the key witnesses who expressed reservations about the proposed legislation. Ruckleshaus indicated that drinking water was a top priority at the EPA; the agency had recently increased its research activities and created an interdisciplinary staff team to review existing federal standards. Although Ruckleshaus was generally supportive of the idea of federal legislation, he seemed to believe that more time was needed to determine the proper role for the federal government (*Safe Drinking Water*, 1971).

Likewise, some drinking water trade associations were reluctant to fully support the legislation. The American Water Works Association (AWWA) expressed concern that mandatory federal standards would prove costly and reduce innovation. Of particular concern to the AWWA were the costs that meeting federal mandates would impose on smaller water systems. A federal mandate would presumably require smaller water systems to invest in treatment technologies they were not currently using and might not be able to afford. The AWWA welcomed increased federal involvement in areas such as research and technical training, but seemed more comfortable with the existing system of voluntary standards that gave states and individual water systems broad latitude to experiment with treatment technologies and enforcement mechanisms (*Safe Drinking Water,* 1971).

AWWA's concerns were partially mitigated when the EPA outlined a legislative proposal that provided for mandatory national standards, but left the states in charge of enforcement (*Potable Waters,* 1972). Provision for state enforcement was an important political selling point in two respects. First, it more closely resembled the status quo to which state regulators and the drinking water industry were accustomed. Second, the proposed federal/state division of authority over drinking water resembled provisions contained in the recently passed Clean Air Act and Federal Water Pollution Control Act Amendments. Both of these pieces of legislation received strong bipartisan support in Congress, so emulating their basic regulatory structure in the drinking water bill seemed like a safe political bet.

The EPA's legislative recommendations received cautious but unified support among groups affiliated with the drinking water industry (*Safe Drinking Water Act,* 1973). Nevertheless, it took until nearly the end of 1974 to pass safe drinking water legislation for a number of reasons. During 1972, the relevant Senate and House committees were busy working on the rather expansive Federal Water Pollution Control Act Amendments, which meant that consideration of safe drinking water legislation was essentially stalled until early 1973. Furthermore, once the Water Pollution Control Amendments were passed, there didn't seem to be a lot of enthusiasm on the part of President Nixon or many members of Congress to tackle another piece of water legislation. Nevertheless, advocates for drinking water regulation managed to keep the issue on the agenda. A General Accounting Office (GAO) study of 446 water systems in six states found that bacterial contamination of drinking water remained a serious public health threat (US General Accounting Office [GAO], 1973). These findings served as the basis for additional congressional hearings that helped build support for legislative action. More importantly, however, the studies linking drinking water contamination to cancer caught the attention of both legislators and the general public. As Rep. Rogers would later remark, these findings helped advocates of drinking water legislation obtain "the visibility and sense of urgency which [they] needed" (Russell, 1978, p. 5).

With little time left before the end of the legislative session, the Safe Drinking Water Act was signed into law by President Ford on December 16, 1974. That "sense of urgency" of which Rep. Rogers spoke was clearly reflected in the provisions of the new law, which required the EPA to quickly write a comprehensive set of standards known as the National Primary Drinking Water Regulations (NPDWRs). The EPA was instructed to promulgate an interim set of regulations within 180 days, though Congress anticipated that agency could complete this task by simply ratifying the existing PHS standards. In the long run, however, Congress anticipated a broader, more ambitious set of standards. Money was appropriated for the National Academy of Sciences (NAS) to conduct an extensive study of drinking water contaminants that would serve as the basis for formulating more permanent standards. No more than 100 days following receipt of the NAS report, the EPA was statutorily required to promulgate a permanent set of NPDWRs (Safe Drinking Water Act [SDWA] of 1974). Once promulgated, the NPDWRs became binding standards enforceable against public water systems across the US.

A careful reading of the SDWA reveals that Congress intended the process of creating the NPDWRs to be scientifically rigorous. In addition to the NAS study, Congress created the National Drinking Water Advisory Council (NDWAC), a panel of citizens and external experts that was given a broad mandate to "advise, consult with, and make recommendations" to the EPA concerning drinking water policy (SDWA of 1974, p. 28). To further assist in the development and implementation of standards, the EPA was required to conduct nine additional studies related to the detection and prevention of contamination in public water supplies. Taken together, the ambitious array of research activities authorized by Congress was meant to ensure that the EPA would fulfill the statutory goal of identifying drinking water conditions under which "no known or anticipated adverse effects on the health of persons occur" (p. 3).

Beyond this statutorily mandated scientific rigor, the standard-setting process was mainly a continuation of established PHS practices with some new elements. On the one hand, Congress anticipated that the NPDWRs would mainly be quantitative pollution limits similar in form to the PHS standards. The main difference was that under the new law these quantitative limits were given a new name: Maximum Contaminant Levels (MCLs). Indeed, the standard-setting process that Congress designed under the new law seemed to anticipate that the old PHS standards would serve as a foundation upon which the EPA would build. At the same time, however, Congress took the somewhat novel step of allowing the EPA to set technology standards, rather than MCLs, for some pollutants. In other words, rather than designate a quantitative limit on a particular pollutant, the EPA might instead simply mandate that public water systems employ a particular type of pollution control technology that reduced contamination to roughly acceptable levels (SDWA of 1974).

The decision to authorize federal technology standards made a certain amount of practical sense. Congress seemed to understand that it might be scientifically very difficult to identify quantitative "safe" levels for many pollutants. Conversely, it might be scientifically possible to estimate a safe levels of exposure for humans, but technologically impossible to attain those pollution reductions. In either instance, it might simply be more practical to just mandate a technology that made the finished water as clean as possible. From a political standpoint, however, federal technology standards went against established notions of federal roles and responsibilities in drinking water regulation. Under the old PHS standard-setting regime, the federal government set standards, but states were left free to choose water sources and treatment technologies. This arrangement was necessary in part because it respected the traditional authority of states to manage local health and safety issues, but also because individual states and localities could not easily adhere to a one-size-fits-all technological solution. Smaller water systems simply could not afford the same range of treatment technologies as larger systems without passing along unreasonable costs to consumers. Therefore, extensive deployment of federal technology standards would likely result in economic hardship for some water providers, and new forms of political conflict within the drinking water policy community.

In an effort to counteract these potential political pitfalls, Congress wrote a requirement into the law that all NPDWRs ultimately be based on what was "feasible" to protect the public health. In this instance, feasibility was defined in terms of the "best technology, treatment techniques, and other means, which the Administrator finds are generally available (taking cost into consideration)" (SDWA of 1974, p. 4). The implication seemed to be that EPA should limit itself to prescribing treatment technologies that were readily available and did not impose an excessive cost burden on the water industry. Whether or not the EPA and water providers could easily agree which technologies were "generally available" and cost effective, however, would not be known until the implementation process was underway.

IMPLEMENTING THE SDWA

During the first year of SDWA implementation activities, the EPA got off to a fast start. The Interim NPDWRs were quickly adopted, which allowed the agency to begin making preparations for the more permanent standards that would need to be adopted once the NAS study was delivered. In particular, the EPA began to consider the plight of small drinking water systems that would incur higher costs once the new standards were in place. The EPA anticipated that it could use short-term mechanisms such as variances and exemptions allowed under the law to help defray the costs, though the long-term solution would require the agency to carefully consider costs when setting drinking water standards (National Interim Primary Drinking Water Standards, 1975).

No formal action could be taken on the final standards until the NAS delivered its much-anticipated review of the available scientific evidence concerning known drinking water contaminants. In the mean time, however, the EPA undertook its own examination of what it considered to be the most significant area where new NPDWRs were needed: organic contaminants. By that time, the EPA understood that there were hundreds of distinct organic contaminants emanating from multiple sources. On the one hand, you had chloroform and other trihalomethanes, which were a byproduct of the chlorination process. Because disinfection by chlorination was so widely used, trihalomethanes were virtually ubiquitous in drinking water, and occurred in larger concentrations than other organic contaminants. At the same time, however, there were potentially hundreds of manmade "synthetic organics" entering drinking water sources in smaller quantities, but from a diverse and difficult to identify array of pollution sources (*Safe Drinking Water Act Amendments*, 1977).

At least some of these organic contaminants were suspected to be carcinogenic in humans, though the scientific data was far from complete.

Following a review of the available scientific evidence by the EPA's Science Advisory Board, the agency was forced to conclude that most available scientific studies suffered from inappropriate or flawed research designs. The best the EPA could conclude from these studies was that there were likely health risks associated with some of the organic contaminants that could not easily be quantified. Nevertheless, the EPA concluded that the potential health risks and general pervasiveness of these chemicals required them to take action. Setting MCLs for organic contaminants was simply not practical, given both the sheer number of such chemicals and the lack of good information about them. Ultimately, the EPA decided to identify and mandate treatment technologies capable of mitigating organic contaminants to the maximum extent feasible (Organic Chemical Contaminants, 1976).

During the late 1970s, two events partially derailed the EPA's implementation of the SDWA. The first event was the release of the long-awaited NAS report on drinking water contaminants. The report was finally released in early 1977, after roughly two years of thoughtful review of the available scientific evidence. Although the report did an adequate job of summarizing existing scientific knowledge, it did not provide what Congress and the EPA had anticipated: a comprehensive set of quantified "safe" levels of exposure to drinking water pollutants that could easily be translated into MCLs (National Research Council, 1977). Speaking before Congress the following year, one high-ranking EPA official explained that implementation delays would be inevitable because the NAS study "did not provide the data and guidance the agency expected" and the agency would therefore need to "conduct additional evaluation as part of the process of developing more comprehensive standards" (*Safe Drinking Water Act Oversight*, 1978, p. 6). The implications of this statement were clear: the EPA had no hope of completing the permanent drinking water standards by the statutorily mandated deadline.

The second event was the EPA's decision to mandate granular activated carbon (GAC) as the treatment technology for organic contaminants. GAC was by no means a proven or cheap solution, but it was the best idea that the EPA had at the time. Compliance costs for public water systems would include hundreds of millions of dollars in upfront capital expenditures, and tens of millions in annual operation and maintenance costs (Control of Organic Chemical Contaminants, 1978). The average cost per family would likely range from three to sixteen dollars annually, depending on the number of homes served by the water system. At the same time, nobody could really quantify just how much safer drinking water would be after GAC treatment. One set of authors who were close to the EPA decision-making process summed it up well:

> The problem is that no one knows what levels of which chemicals would be removed by granular activated carbon. There are many thousands of organic chemicals in drinking water … [and] only a few have been tested for carcinogenicity in the usual animal tests … Yet the value of granular activated carbon is that it removes a wide variety of these compounds from water. The agency has suggested that it makes sense to spend a nominal amount … to take advantage of a known technology … Yet neither the agency nor its critics has been able to develop a method of analysis that would help determine whether or not granular activated carbon is worth its costs (Kimm, Kuzmack, and Schnare, 1981, p. 241).

Recognizing its inability to provide adequate cost justifications for GAC, the EPA nevertheless chose to mandate the treatment technology for most public water systems.

Over the next several years, the GAC mandate proved to be the most politically divisive issue in the history of federal drinking water policy up to that time. A very public water industry backlash against the new rules played out as a series of sometimes animated exchanges during congressional hearings. Some critics questioned the science behind the EPA's decision. For instance, a representative of the National Association of Water Companies characterized the health risks associated with some organics as "miniscule … and poorly proven," and the GAC treatment technology as "uncertain" and likely to cause health hazards of its own (*Federal Safe Drinking Water Act*, 1978, pp. 3-4). In the same hearing, a representative of the Coalition for Safe Drinking Water —another industry trade association—suggested that the EPA should be required to conduct extensive demonstration projects to show that GAC technology was both feasible and effective.

For many in the water industry, however, the real issue seemed to be that the EPA had overstepped its bounds by mandating a treatment technology. Water industry groups portrayed themselves as the real technical experts, who were best prepared to meet the day-to-day safety challenges associated with providing safe drinking water. As a representative of the National Association of Water Companies noted during testimony before Congress:

> It is with all due respect we contend EPA will never possess the practical knowledge equal to that which is available to water supply professionals. Much more can be accomplished by devoting the tremendous resources EPA possesses toward the scientific definition of appropriate water quality which will not pose an unreasonable risk to human health (*Safe Drinking Water*, 1982, p.4).

In other words, the EPA would do well to stick to the traditional federal role of setting quantitative, science-based pollution limitations, while leaving matters involving technology and implementation to the states and to water professionals.

CONCLUSION

The SDWA was passed in 1974 with good intentions, but with only a rudimentary understanding of what it would take to clean up the nation's drinking water. There remained large gaps in scientific knowledge concerning both the range of contaminants occurring in US drinking water and the health effects associated with them. Furthermore, Congress underestimated the costs associated with mitigating chemical contaminants. By the early 1980s, slow implementation of the SDWA combined with growing political tensions between the EPA and the water industry caused some members of Congress to conclude that further legislative action was required.

Chapter 4

POLITICAL CONFLICT AND POLICY REFORM

In the early 1980s, Congress began considering proposals to substantially reform the SDWA. Less than stellar implementation of the law by the EPA and the states provided the impetus for legislative action. As legislators and their staff went about the business of crafting new legislation, they drew inspiration in part from ideas associated with what Eisner (2000) calls the "efficiency regime." By the end of the 1970s, the regulatory zeal that fueled passage of the new social regulation gave way to calls for either deregulation or more efficient regulation. Originating in economic studies published in the 1960s and 1970s, this new gospel of efficiency was taken up by interest groups and political leaders from across the ideological spectrum. Driven by the forces of electoral competition, candidates and leaders of all political stripes were looking to advance their careers by touting their own proposals for deregulation or regulatory reform (Derthick and Quirk, 1985).

Within the context of drinking water regulation, however, the efficiency regime ran head on into long established notions of what was needed to protect the public health. As legislative leaders began to seriously consider amendments to the SDWA, they found it difficult to conceptualize a policy regime in which economic efficiency and protection of the public health were both equally attainable. As a result, amendments to the SDWA passed in the 1980s were at best small departures from existing policy arrangements.

Over time, however, forces within the environmental policy bureaucracy reconciled the competing values of efficiency and public health protection by advocating a new philosophy known as "risk management." Advocates of risk management believed that regulators could use the tools of risk assessment to scientifically determine which contaminants posed the greatest risks to human health and the environment. Once this process was complete, regulators could then work with industry and citizen groups to decide which risks deserved immediate attention, and which ones could be tolerated for the foreseeable future (Habicht, 1994). As the 1980s came to a close, risk management advocates inside the EPA argued that these principles should be the guiding intellectual force in federal drinking water regulation.

This chapter chronicles these developments as they unfolded throughout much of the 1980s, starting with the emergence of the efficiency regime, and ending with the development of risk management principles.

THE EFFICIENCY REGIME

During the late 1970s the Washington community considered the possibility that the extensive array of regulatory laws passed during the preceding century imposed excessive financial burdens on American corporations and consumers. The major factor precipitating this change of heart was the end of post-World War II economic prosperity. By the mid-1970s, the appearance of "stagflation"—simultaneously high unemployment and inflation—caused traditional Keynesian theories of economic policy to fall into disrepute. As Washington officials sought answers to the nation's economic dilemmas, they were drawn to new policy ideas that preached the need to reform the nation's inefficient and sometimes outdated regulatory policies (Derthick and Quirk, 1985; Eisner, 2000).

By that time, federal regulation was the universal scapegoat of interest groups and intellectuals on both the left and right ends of the political spectrum. On the left, consumer advocates like Ralph Nader and his associates argued that regulation of the transportation sector served mainly to increase the profits of large corporations while imposing unnecessary costs on consumers (Green, 1973). On the right, economists writing in the Chicago School tradition argued that regulation created marketplace inefficiencies and imposed excessive compliance costs on businesses. Murray Weidenbaum (1978), for instance, estimated that the annual costs associated with federal regulation would exceed $100 billion by 1980. These critiques caught the attention of both Republicans and Democrats in Congress, and helped put deregulation on the legislative agenda (Derthick and Quirk, 1985).

Perhaps the most significant deregulatory initiative passed during this time was the Airline Deregulation Act of 1978, which eliminated all price and entry regulation of the commercial aviation industry (Brown, 1987). In most instances, however, deregulation was carried out through the appointment of high-level agency officials who used delegated statutory authority to repeal burdensome regulations (Derthick and Quirk, 1985). Both presidents Carter and Reagan, for instance, appointed Federal Communications Commission chairmen that pursued deregulatory agendas (Zarkin, 2003). Guided by economic analysis, deregulation-minded agency officials sought to eliminate regulations that did not produce substantial and quantifiable social benefits.

Deregulation did not substantially affect the EPA, however, for two overarching reasons. First, environmental laws remained fairly popular with the American public. Second, federal pollution control laws like the SDWA did not give administrators the discretion to chose to deregulate. The EPA, however, was more directly impacted by a related trend known as "regulatory reform" (Harris and Milkis, 1996). Regulatory reform was essentially the idea that the application of economic analysis could be used to make existing regulations more efficient. In particular, cost-benefit analysis was promoted as the best methodology to help regulatory agencies choose only those regulations that offered substantial social benefits, given both the costs of compliance and administration.

Both Presidents Carter and Reagan appointed EPA administrators who advocated regulatory reform. During the tenure of Douglas Costle (1977-1981), the EPA's Office of Planning and Management became the major locus of economic analysis within the agency (Cook, 1988). It was during the tenure of Administrator Anne Gorsuch Buford (1981-83), however, that regulatory reform was taken too far. Buford's attempts to cut the agency's budget and reorganize its personnel were viewed by members of Congress as efforts to impede the implementation of the nation's environmental laws (Harris and Milkis, 1996). Buford underestimated public support for national laws protecting the environmental and public health. When she left office, the EPA's public image was in decline and an impatient Congress was anxious to get the nation's environmental protection program back on track.

DRINKING WATER RETURNS TO THE AGENDA

At the same time that the newfound zeal for economic efficiency brought new ideas into the policy process, a more specific series of factors brought drinking water issues back onto the legislative agenda. One such factor was a lack of enforcement by state governments. In 1980, the EPA recorded 146,000 violations of its drinking water standards across 28,000 of the nation's public water supplies. Problems with enforcement were mainly a function of cost. Roughly 13,600 water systems needed capital improvements to be able to comply with all federal drinking water standards. Furthermore, many of the smaller water systems lacked the full-time, trained staff needed to ensure drinking water quality. Recognizing the need for increased federal guidance, in 1981 the EPA began working more closely with states to improve their compliance plans. A GAO (1982) study undertaken the following year, however, found that many state implementation strategies remained deficient.

While members of Congress were concerned by the mounting costs and implementation challenges faced by the states, they were also frustrated by the EPA's slow and seemingly incompetent implementation of the SDWA. Congress originally anticipated that the EPA would have a permanent set of NPDWRs in place by 1977. Five years later, however, final regulations were nowhere in sight. Furthermore, in 1983 the EPA made the controversial decision to repeal the GAC treatment technology mandate. Relying heavily on cost-benefit analysis the EPA reversed itself, concluding that other technologies such as changes in the water disinfection process were generally available and cheaper than GAC (National Interim Primary Drinking Water Regulations: Trihalomethanes, 1983). The EPA seemed to forget, however, that alternative disinfection processes would only deal with trihalomethanes, whereas GAC technology held the potential to mitigate a broader array of organic contaminants.

Congress was also frustrated by the EPA's slow progress toward setting an appropriate standard for the control of lead contamination in drinking water. Admittedly, lead contamination presented federal regulators with a pernicious problem. Lead places humans at risk for a host of health problems including high blood pressure, impaired reproductive systems, and reduced cognitive development in children. Furthermore, lead is more difficult to treat than many other drinking water contaminants. Most contaminants enter water at the source and can be filtered out by system operators at the plant. Lead, however, enters water as it traveled through pipes on its way to the consumer's tap, making mitigation of the problem costly and difficult (EPA, 1986). EPA staff initially recommended a technology standard that combined anti-corrosion measures with partial replacement of the lead pipe infrastructure. The EPA estimated that these measures would reduce lead contamination to roughly ten micrograms per liter of water—a level that was considered feasible, though not necessarily ideal. The program, however, required \$3 billion in capital outlays, with an added \$1 billion in annual maintenance costs. The EPA later concluded that the benefits did not justify the costs, and revised the recommendations to include only the less expensive anti-corrosion measures (*Lead Contamination,* 1988).

In many respects, organic contaminants and lead contamination epitomized the struggles that the EPA and Congress faced as they attempted to forge a workable path in drinking water regulation. On the one hand, the traditional approach to drinking water standards—the approach pioneered by the PHS many decades earlier—emphasized public health protection as the preeminent value in drinking water regulation. On the other hand, federal mandates placed a growing financial burden on the water industry and state governments that could not be ignored. Therefore, if legislators were going to agree on a package of revisions to the SDWA, they would first need to agree on the appropriate balance between health protection and costs.

LEGISLATIVE DELIBERATIONS

Legislative deliberations effectively commenced in late 1981, when Senator Slade Gorton (R-WA), Chair of the Senate Subcommittee on Toxic Substances and Environmental Oversight, introduced S. 1866—The Safe Drinking Water Regulatory Reform Act. As the name of the bill suggests, Gorton clearly came down on the side of a regulatory reform approach to drinking water regulation. S. 1866 proposed to place an onerous new set of requirements on the EPA, including a requirement that drinking water standards be accompanied by a formal cost-benefit analysis. Additionally, the bill would have required the EPA to provide an elaborate statement explaining why any alternative regulatory proposals were rejected "taking into account effectiveness, efficiency, and costs" (*Safe Drinking Water,* 1982, p. 223).

Perhaps more significant, however, was a requirement in the bill that the EPA only regulate drinking water contaminants when they posed an "unreasonable risk" to human health. By that time, the unreasonable risk standard was firmly established in federal environmental law, having already been incorporated into the Toxic Substances Control Act of 1976. Environmentalists criticized the unreasonable risk standard because it placed a high burden of proof on the EPA to show that the health risks associated with a contaminant were in fact "unreasonable." Indeed, by the early 1980s the EPA had made little progress in regulating potentially toxic chemicals under the Toxic Substances Control Act (GAO, 1990b). It therefore stood to reason that the unreasonable risk standard would only further slow the EPA's efforts to regulate drinking water contaminants.

During hearings held in 1982, S. 1866 received moderate support from both the water industry and state regulators. Current EPA officials expressed support specifically for the requirement that cost-benefit analysis be incorporated into rulemaking proceedings. Former EPA officials, however, greeted the bill less favorably. Thomas Jorling, Assistant EPA Administrator during the Carter Administration, raised concerns that the unreasonable risk standard would erode the "preventative principle" that had served as the cornerstone of drinking water regulation for many decades (*Safe Drinking Water,* 1982, p. 58). Jorling also suggested that the standard would force regulators into the rather callous business of attempting to place a monetary value on human life in order to justify regulations.

Robert Neal, Chair of the NDWAC, attempted a rejoinder by suggesting that the language in the legislation be changed from "unreasonable risk" to "unacceptable risk," which he defined as "an unacceptable level of risk … based on sound scientific information on the nature of the risk and taking costs into consideration" (*Safe Drinking Water,* 1982, p. 251). By this time, however, witnesses were mainly arguing over semantics and seemed to be creating confusion rather than bringing clarity to the issue. By the end of the hearing, Senator Gorton was ready to abandon the legislation altogether, concluding that "the public health will not be improved by changing the present language in the act" (*Safe Drinking Water,* 1982, p. 543). As a result, legislative revision of drinking water policy remained more or less in discussion mode for the next three years.

The prospect of comprehensive legislative reform resurfaced once again, however, when Senator David Durenberger (R-MN) ascended to the chairmanship of the Subcommittee on Toxic Substances and Environmental Oversight. Unlike Gorton, Durenberger brought to the debate a much more preventative posture that prioritized protection of the public health over costs. Durenberger proposed a bill that was intended to jump start implementation of the SDWA by creating a very rigid set of timetables and standard setting requirements for the EPA. For Durenberger, the reasons for proposing such preventative legislation were highly personal, as was revealed in an exchange between him and Assistant EPA Administrator Jack Ravan in a Senate hearing:

> Mr. Ravan: … I am not certain that increased rigidity by statute, upon the [EPA] Administrator, solves the problem, given what you and I both know to be the circumstances, which is in fact, we are now living in the chemical age … I doubt, Mr. Chairman, there will ever come an annual sequence of hearings, in fact, where you and I will not both have discovered some new threat, that we didn't know about yesterday, that now we have to do something about, under a very cumbersome process. The thing that I would ask you for is recognition of that process, rather than taking away whatever flexibility that the Administrator might have to deal with the science and the toxicology available to us on the substances as they do occur in drinking water.
>
> Sen. Durenberger: Jack, I find myself in agreement with you, but I am increasingly troubled by the fact that last week a very, very close friend's 26-year old son was diagnosed as having a massive cancer. All that did was remind me … that my first wife went the same way. I am frankly scared to death about the chemicals that are out there and that we are consuming in one way or another. I am also bothered by the fact that with this large deficit that we have constructed for ourselves … there are people in this government that believe that we shouldn't … put money into all this analysis because we do not have it … So … I am begging for something more than 'let us keep the discretion to continue to do things as usual …' to come from the Environmental Protection Agency (*Safe Drinking Water Act Amendments,* 1985, pp. 13-14).

In spite of objections from the EPA, the water industry, and state governments, Durenberger's preventative approach to drinking water regulation prevailed.

The Safe Drinking Water Act Amendments of 1986 passed the Senate by a vote of 94-0, and the House by a vote of 382-21 (Ketcham-Colwell, 1986). The new law required the EPA to promulgate revised drinking water regulations for a list of eighty-three contaminants within a strict timeframe of three years. During that time, if the EPA concluded that the regulation of substances not contained on the list was "more likely to be protective of public health," the Administrator could choose to make a maximum of seven substitutions (SDWA Amendments of 1986, p. 2). The Amendments also specified that by January 1st, 1988 and in three-year intervals thereafter, the EPA was required to work with a special advisory board to compile a list of additional contaminants that constituted candidates for regulation. Within each three-year interval, the EPA was required to regulate at least twenty-five contaminants contained on the list. Finally, the 1986 Amendments also changed the definition of a "feasible" treatment technology from one that was "generally available" to the "best available," but still required the EPA to take costs into consideration. The Amendments also overturned the EPA's decision not to recognize GAC as a "feasible" technology for purposes of controlling synthetic organic contaminants (SDWA Amendments of 1986, pp. 3-4).

Two years later, concerns that children were being exposed to too much lead in drinking water caused Congress to pass the Lead Contamination Control Act. The law required all water coolers sold in interstate commerce in the US to be "lead free" as defined by the statute. The EPA was also required to develop a testing program to help schools identify lead contamination in drinking fountains. In addition, $90 million in grant money was made available over a three-year period to help schools identify and remedy lead contamination in drinking water (Lead Contamination Control Act of 1988).

The Emergence of "Risk Management" Principles

With the passage of the SDWA Amendments of 1986, Congress appeared to ignore calls from both the EPA and the water industry for a better balance between costs and public health considerations in the creation of national drinking water standards. Moreover, the 1986 amendments and the Lead Contamination Control Act, taken together, were a break from past policy practice in that they established a clear, though still limited, role for the federal government in setting drinking water technology standards. In this respect, the SDWA Amendments of the 1980s, much like the original 1974 enactment, represented piecemeal breaks away from the old drinking water policy regime that existed prior to 1974.

At the same time, however, the 1980s amendments were also a continuation of past policy practices. The ideas associated with the "efficiency regime" helped legislators like Senator Gorton craft legislation that emphasized economic efficiency, but did not provide a comprehensive vision of how public health protection and cost containment could function as complimentary goals. For this reason, when it came time to formulate revisions to the SDWA, legislators were drawn back to the precautionary approach that had prevailed for decades. Nevertheless, the legislative decision to ignore the lingering tension between economic efficiency and health considerations did not make these issues go away. When former EPA Administrator William Ruckleshaus was asked by President Reagan to return to that position in 1983, he made an effort to change the conversation by framing cost and environmental quality as compatible principles. These efforts eventually led to the ascendance of the risk management policy paradigm.

The risk management paradigm came to prominence within the US environmental policy community as a result of an article Ruckleshaus published as he prepared to leave office in 1985. Reflecting on his most recent experiences as EPA Administrator, Ruckleshaus concluded that the regulatory philosophy underlying the environmental policy regime of the 1970s was becoming outmoded. Regulators were unduly focused on mitigating poorly understood and, in some cases, minute carcinogenic risks without regard to cost or the magnitude of the threat. Instead, Ruckleshaus believed that the EPA needed to set more effective environmental policy priorities in a world characterized by scarce resources. This was the essence of what Ruckleshaus called "risk management" (Ruckleshaus, 1985).

Ruckleshaus vaguely stated that risk management involved "adjusting our environmental policies to attain the array of social goods … that forms our vision of how we want the world to be" (Ruckleshaus, 1985, p. 31).

Ruckleshaus believed that science and economics were valuable tools that could be employed to help society arrive at optimal environmental policy priorities. Ultimately, however, legitimate environmental policies could only be fashioned in cooperation with an informed, engaged citizenry. Therefore, under the risk management paradigm regulators have an obligation to communicate effectively with the public concerning the nature of risks and the rationale behind policy decisions. As Ruckleshaus put it "transparency is the object of the whole process, and public trust is the ultimate goal" (p. 35).

Lee Thomas, Ruckleshaus' successor as EPA Administrator, began the process of systematically putting risk management principles into practice. For Thomas, the formulation of sound risk priorities first required a systematic comparison of the risks posed by various environmental problems. To this end, the EPA assembled a series of working groups and undertook an ambitious "comparative risk" study that examined and ranked thirty-one environmental problems against four distinct categories of risk: cancer risks, non-cancer health risks, ecological risks, and welfare risks. The study concluded with the promulgation of a report entitled *Unfinished Business* that provided the EPA with a set of science-based environmental policy priorities intended to inform future policy and budgetary planning decisions (EPA, 1987).

However well-intentioned the EPAs comparative risk project might have been, science-based priorities proved difficult to implement for two main reasons. First, existing legal arrangements did not easily allow for changes in environmental policy priorities. Environmental statutes like the SDWA Amendments of 1986 and the Water Quality Act of 1987 required the EPA to regulate specific pollutants according to fairly rigid timetables. Furthermore, the most significant risks identified by the EPA in the comparative risk study were not the American public's biggest environmental concerns. For instance, whereas the EPA concluded that indoor air pollution, global warming, and consumer products exposure presented some of the most significant environmental risks, public opinion polls showed that Americans were more concerned with lower risk problems like chemical waste disposal, which had received more media attention in recent years. These realities suggested that science-based risk priorities could only be realized through a long-run strategy of dialog involving regulators, politicians, and citizens (Fiorino, 1990).

In spite of these challenges, thought leaders in and around the EPA remained committed to a long-run strategy of implementing risk management principles. The EPA's Science Advisory Board (1990) issued a report entitled *Reducing Risk* in which it highlighted the need for policy reforms rooted in the risk management paradigm. The report was originally intended to be little more than an external review of the scientific conclusions contained in the *Unfinished Business* report. The text of *Reducing Risk*, however, read more like a policy position paper than a set of scientific findings. In particular, the report was structured around six recommendations that built on the risk management principles articulated by Ruckleshaus five years earlier: 1) target environmental protection efforts toward opportunities for the greatest risk reduction, 2) work to improve the use of scientific risk assessment, 3) reflect risk-based priorities in strategic planning and budgeting decisions, 4) make pollution prevention the preferred option for reducing risk, 5) work to improve public understanding of environmental risks, and 6) better account for environmental effects in economic analyses (EPA, 1990, p. 6). Drinking water was among the areas of environmental and public health policy identified in *Reducing Risk* as deserving of particular attention by the EPA.

CONCLUSION

During the 1980s, legislative action once again led to piecemeal changes in the drinking water policy regime. The 1986 and 1988 amendments further strengthened the federal role in drinking water regulation, but did not resolve a key issue that divided the federal and state governments: the appropriate balance between public health protection and economic costs. By the end of the 1980s, the risk management paradigm promised to help bridge this gap by offering an intellectual roadmap toward more balanced environmental and public health policies.

Depending on your perspective, risk management principles could either be viewed as a radical departure or simply more of the same. On the one hand, the language contained in the SDWA always presupposed some balance between costs and public health protection. Therefore, advocates of the status quo could be reassured that risk management proponents were simply proposing more specific strategies that reinforced existing SDWA principles and values. On the other hand, neither the SDWA nor the "traditional" pre-1970 approach to drinking water regulation presupposed that it would be necessary to make economic trade-offs or ask the public to tolerate some risks in order to protect the overall public health. When viewed this way, risk management principles could be characterized as a true paradigm shift that promised to take drinking water regulation in a new direction. The beauty of risk management was that it framed drinking water issues in a way that seemed to offer something for everyone. Ultimately, however, the success of risk management principles depended on whether or not they could be fashioned into specific legislative proposals that garnered the support of legislators and key stakeholders in the drinking water policy community.

Chapter 5

THE TRIUMPH OF RISK MANAGEMENT, 1990-2015

In the spring of 1993, the municipal water supply in Milwaukee, WI was infected with Cryptosporidium—a waterborne parasite that rapidly caused roughly 400,000 people to experience severe gastrointestinal distress (Matukaitis, 1997). Milwaukee was not the first Cryptosporidium outbreak, but it was by far the largest up to that time; it caught the attention of the national media, and caused the public to become concerned about the state of America's drinking water. For members of Congress, however, the Milwaukee crisis only brought into sharp relief what they already knew: state enforcement of the SDWA remained inadequate. In the case of Cryptosporidium, state regulators were generally not equipped to perform the kind of laboratory testing needed to identify the parasite (Altman, 1993; Pontius, 1993). Furthermore, a series of GAO reports in the early 1990s indicated a bigger problem: many states remained unwilling or unable to invest the resources to conduct inspections and properly treat water. Subsequent congressional hearings paved the way for yet another set of major revisions to the SDWA in 1996.

This time, however, members of Congress were willing to confront the problem they had seemingly ignored in 1986: the need to find a workable balance between public health protection and the economic costs associated with regulation. One factor contributing to this change in thinking was undoubtedly the election of a conservative, Republican majority to Congress in 1994. The Republican Contract with America included a pledge to reduce regulation and balance the federal budget: values that certainly counseled in favor of a new approach to drinking water regulation. The SDWA Amendments of 1996, however, passed Congress with large bipartisan majorities, indicating that additional factors were driving policy change. In this chapter, it is argued that passage of the SDWA Amendments of 1996 was most directly enabled by new policy ideas rooted in the risk management paradigm. As will be demonstrated in the pages that follow, risk management provided both specific policy solutions and a new way of talking about drinking water issues that facilitated passage of the new law.

The remainder of this chapter is dedicated to a discussion of the passage and implementation of the SDWA Amendments of 1996. The next section chronicles the legislative debates that ultimately led to the adoption of risk management principles by Congress. The remaining sections describe how risk management principles were institutionalized through EPA implementation activities between 1996 and 2015.

Legislative Action Renewed

In the early 1990s, Congress once again began to consider amendments to the SDWA after a series of GAO reports uncovered ongoing deficiencies in state-level enforcement activities. In particular, the GAO (1990a) examined ninety-five major instances of SDWA non-compliance across six states, and found that state enforcement officials took action in only twenty-four of those cases. Furthermore, forty-six of the ninety-five cases remained persistent non-compliance problems for four years or longer. Neither the EPA nor state enforcement officials denied these findings, but both indicated that a lack of funding, rather than a lack of will, was the real reason for non-enforcement. The EPA claimed that new regulatory requirements mandated by the SDWA Amendments of 1986 would cost the states as much as $3 billion per year for the next twenty years. Many state governments, however, did not even employ the personnel or laboratory facilities needed to implement existing SDWA requirements (GAO, 1992). The EPA was legally required to withdraw enforcement responsibilities from states that demonstrated severe deficiencies in their SDWA implementation programs. Doing so, however, would not improve SDWA implementation because the EPA lacked the resources to unilaterally administer the law (GAO, 1993).

The message to members of Congress was clear: either appropriate large sums of money to assist states with their implementation responsibilities, or find a way to protect the public health at a lower cost. Finding billions of dollars in new funding for environmental programs was politically difficult at a time when the federal government perpetually ran a budget deficit. The prospect of significantly increasing SDWA funding was further decreased when the Republican Party won control of Congress in 1994 on a platform that included balancing the federal budget and reducing costs associated with federal mandates to the states. In a political environment characterized by limited resources and growing hostility to excessive federal intervention, public health-minded members of both political parties in Congress understood the need to find a fiscally responsible way to protect the public.

This political context created a window of opportunity for EPA officials to advance policy reform proposals rooted in risk management principles. In a set of policy recommendations announced in the fall of 1993, the EPA was careful to address the cost concerns advanced by state and local governments. The EPA proposed a scheme for providing long-term financial assistance in the form of grants that would allow states to set up revolving loan funds to assist local water systems. At the same time, however, the EPA made it clear that cost containment also required better priority setting that targeted regulatory resources to the most dangerous pollutants. To this end, the EPA requested that Congress replace the rigid 1986 regulatory requirements with "a more scientific procedure for determining which contaminants EPA and the states should regulate" (EPA, 1993, p. 1). In addition, the EPA proposed that costs could be reduced in the long run through pollution prevention measures such as a source water protection program. In a more elaborate set of recommendations issued in early 1995, the EPA reiterated many of these earlier principles while also stressing that better priority setting required public education because "an understanding of drinking water conditions helps rate-payers recognize the need for problem-solving actions by their community system or state" (EPA, 1995, p. 10).

Legislation introduced around that time incorporated key elements of the EPA's risk management proposals. Sen. Max Baucus (D-MT) introduced legislation that he characterized as "returning scientific judgment to the [EPA] Administrator in deciding whether to regulate future contaminants" (*Safe Drinking Water Act Amendments*, 1993, p. 2). The central feature of the Baucus bill was a requirement that the EPA Administrator set risk-based priorities in the regulation of drinking water contaminants. Baucus made clear his belief that the kind of science-based priority setting he favored could "achieve greater public health protection at lower cost" (p. 3).

The risk-based approach taken by the Baucus bill resonated with a bipartisan group of Senators who were growing frustrated with the status quo in drinking water regulation. In a hearing before the Senate Committee on Environment and Public Works, Senator Mark Hatfield (R-OR) testified about the "enormous sentiment" within his state "that we're spending millions of dollars testing and treating drinking water without relevance to the risk to health" (*Safe Drinking Water Act Amendments*, 1993, p. 4). Senator George Mitchell (D-ME) noted a similar frustration among officials in his state, who were "convinced that the [SDWA] program cost far too much" (p. 5). Senator Dirk Kempthorne (R-ID) noted that citizens and officials in his state believed that the existing SDWA program was "preoccupied with smaller and smaller parts per billion and reducing risks to zero without regard to costs and without regard to benefits" (p. 8). In essence, by the early 1990s, members of Congress were speaking about drinking water issues in a new way. Taken together, the statements indicate that legislators were beginning to embrace a key tenet of risk management: cost and health protection need not be considered opposing values when regulators use science-based priority setting to deal with the most serious risks.

Environmental advocacy groups provided the lone voice in opposition to the risk-based approach to drinking water regulation. In 1993, the NRDC issued a highly publicized series of reports critiquing the quality of US drinking water. The following year, the NRDC issued a second report and joined with a coalition that included five other national environmental advocacy groups to lobby against SDWA reform. With Republicans and Democrats in Congress united behind the risk management approach to SDWA reform, however, NRDC's best efforts were not able to produce more than a delay in the legislation (Pontius, 2003). When the final legislation came up for a vote in the summer of 1996, it passed easily in both houses of Congress.

The SDWA Amendments of 1996 contained provisions that reflected each of the major risk management principles. First, the new law promoted pollution prevention in several ways. The EPA was required to conduct demonstration projects to determine the best mechanisms for assessing and protecting the drinking water sources of large metropolitan areas. In addition, states were required to have a monitoring program for source waters, and encouraged to approve and fund water protection agreements aimed at setting up voluntary partnerships to prevent pollution of drinking water sources (SDWA Amendments of 1996).

The SDWA Amendments also reflected the EPAs desire to set risk-based priorities in the regulation of drinking water contaminants. The rigid requirements contained in the 1986 amendments were replaced with a requirement that the EPA select five new contaminants to be considered for regulation every five years. In selecting contaminants for regulation, the EPA Administrator was instructed to focus on those "contaminants that pose the greatest public health concern" (SDWA Amendments of 1996, p. 7). Whether selecting a contaminant for regulation or setting MCLs, the EPA was required to employ the best peer-reviewed science and to carefully weigh the economic costs against the health benefits accruing from regulation.

Finally, the SDWA amendments contained a range of public notification provisions aimed at educating the public about the quality of drinking water and the efficacy of state and local enforcement. First, local water systems were required to notify consumers of any failure to comply with a federal requirement. In addition, state governments were required to prepare and publish an annual report detailing violations of drinking water regulations by all public water systems within their jurisdiction. Finally, drinking water systems serving more than 10,000 customers were required to prepare and mail to customers an annual consumer confidence report detailing the quality of local drinking water and indicating any health concerns associated with instances of non-compliance (SDWA Amendments of 1996).

IMPLEMENTING RISK MANAGEMENT

The SDWA Amendments of 1996 were a remarkable political achievement at a time when Republicans and Democrats in Congress struggled to find a middle ground on questions relating to environmental protection (Klyza and Sousa, 2008). In this instance, however, bipartisan support coalesced around risk management principles, resulting in a relatively rare achievement: a philosophically consistent and reasonably specific US regulatory statute. Nevertheless, like any major regulatory statute, the real success or failure of the SDWA Amendments depended on how effectively the law was implemented. Given that EPA leaders and staff were among the principal intellectual architects of the risk management philosophy, the agency could certainly be counted on to faithfully execute its new legislative mandate. Effective implementation, however, would not happen over night, because the EPA first needed to develop protocols for the evaluation of the voluminous scientific data needed to effectuate the new risk management strategies.

The most significant risk management component of the SDWA Amendments was a requirement that the EPA engage in risk-based priority setting when considering new contaminants for regulation. Fulfilling this mandate required the EPA to take two steps: compile a Contaminant Candidate List (or CCL), and then select a small subset of contaminants from that list to be considered for regulation. The first CCL needed to be completed within eighteen months after enactment of the SDWA amendments, with subsequent lists following in five-year intervals thereafter. In compiling the CCLs, the EPA is required to identify contaminants that are known or anticipated to occur in public water systems, and which might require regulation. To assist in the creation of the CCLs, the EPA is further required to create and utilize an Unregulated Contaminant Occurrence Database.

Within five years after enactment, and in five-year intervals thereafter, the EPA was also required to select at least five contaminants from the CCL to be considered for regulation – a process more commonly referred to as "regulatory determination." In selecting contaminants for regulatory determination, the EPA Administrator is required to select those that likely pose the greatest risk to human health, and may only impose regulations if a contaminant 1) is likely to have adverse effects on human health, 2) is known to or likely to occur in public water systems at levels likely to pose a risk to humans, and 3) presents a meaningful opportunity for health risk reduction. The EPA is also required to use the best available peer reviewed science to make all new regulatory determinations (SDWA Amendments of 1996).

Implementation of this new legal mandate first required the EPA to undertake the unenviable task of quickly developing objective criteria for determining which among the thousands of known or suspected drinking water contaminants posed the greatest risks to human health. Fortunately, the EPA was not alone in this endeavor; the agency was able to draw upon the input of multiple external advisory groups. For instance, Congress specifically required the EPA to seek advice from its Science Advisory Board (SAB) in the construction of the CCLs (SDWA Amendments of 1996). The SAB had a long history of serving as an external reviewer of the EPA's scientific methods that made it ideally positioned to provide advice concerning the evaluation of CCL contaminant evaluation protocols. Additionally, the SDWA requires the EPA to create and regularly consult the NDWAC. Although the SDWA Amendments of 1996 did not give the NDWAC a specific statutory role with respect to the regulation of new contaminants, its broad mandate to "advise, consult with, and make recommendations" concerning drinking water policy matters virtually guaranteed that it would have a voice in the process (SDWA of 1974). Furthermore, the EPA may request advice from the National Research Council (NRC), which has occasionally provided the EPA with external advice concerning drinking water policy matters, particularly when Congressional appropriations allow such studies to be commissioned (Pontius, 2003).

Even with the assistance of these advisory groups, however, the EPA struggled to develop, refine, and fully implement the needed protocols. To better illustrate this point, the remainder of this chapter chronicles the EPA's implementation of the CCL and regulatory determination requirements between 1996 and 2015.

THE CONTAMINANT CANDIDATE LISTS, 1996-2015

Shortly after passage of the SDWA Amendments, the EPA began work on a contaminant identification protocol for use in the development of the CCLs. In undertaking this task, EPA staff recognized that chemical and microbiological drinking water contaminants each presented distinct scientific challenges. More specifically, the health risks posed by the two categories of contaminants tended to be evaluated through distinct types of scientific evidence, making them difficult to compare along a common scoring mechanism. As such, the EPA proposed to evaluate the two categories of contaminants separately for inclusion on the CCL. Following risk management principles, the EPA proposed that each category be evaluated using a four-step process: 1) identification of contaminants, 2) scientific assessment of health risks, 3) numerical ranking of contaminants based on health risks, and 4) consideration of potential regulatory options that might be available should the contaminant ultimately be regulated (EPA, 1996).

EPA efforts to develop an elaborate contaminant identification methodology, however, were put on hold because the agency needed to comply with the statutory deadline for the completion of the first CCL. With no permanent standardized criteria in place, the EPA worked with the NDWAC and other stakeholders to develop temporary protocols (Announcement of the Draft Drinking Water Contaminant Candidate List, 1997). Nevertheless, time and information constraints forced the EPA to focus on those contaminants for which there was the most available data rather than those that necessarily posed the greatest risk. Working within these limitations, the EPA managed to produce a CCL that contained fifty chemical and ten microbiological contaminants. Among those sixty contaminants, twenty were identified as regulatory determination "priorities" based of the availability of data concerning health effects and occurrence in drinking water (Announcement of the Draft Drinking Water Contaminant Candidate List, 1998).

The EPA was well aware of the deficiencies that permeated the CCL creation process, and committed itself to the development of a "more robust approach to data collection and evaluation" for future CCLs (Announcement of the Draft Drinking Water Contaminant Candidate List, 1998, p. 10285). Over the next several years, the EPA attempted to fulfill this commitment by engaging the NRC and the NDWAC in the development of a better set of protocols. The NRC (2001) subsequently issued a report outlining a two-step process that first involved identifying a "broadly defined universe" of potential drinking water contaminants and narrowing it down to a Preliminary CCL (PCCL) using standardized screening criteria. The second step involved winnowing the PCCL into a final CCL using a numerical algorithm. The NRC also encouraged the EPA to make the CCL development process more transparent by better explaining its decision-making processes and making more information available so that the public could actively participate in the process.

At the EPA's request, The NDWAC reviewed the NRC report and generally approved of the recommendations (National Drinking Water Advisory Council [NDWAC], 2004). The EPA publicly committed itself to implementing the NRC recommendations when it released a draft of the second CCL in 2004. The draft CCL2, however, did not employ the new framework. Instead, the agency simply carried forward the list of contaminants contained on the first CCL (Drinking Water Contaminant Candidate List 2, 2004, 2005). The EPA justified its decision based on the fact that it had further researched many of the chemicals on the first CCL, and was now positioned to determine whether any of those chemicals required regulation. In particular, the EPA set up the Unregulated Contaminant Occurrence Database required by the SDWA and was just beginning to receive reliable data on many of the contaminants contained on the first CCL. Furthermore, the NRC framework required new criteria and decision rules that would take some time to develop. Nevertheless, the EPA used the public announcement as an opportunity to outline the new scheme and request public comment (Drinking Water Contaminant Candidate List 2, 2004).

For the creation of the third CCL in 2008-2009, the EPA developed an elaborate and fairly comprehensive process that incorporated most of the recommendations made by the NRC and NDWAC. The EPA started by working with stakeholder groups to identify a “broad universe” of contaminants through a multi-step process that first involved systematically evaluating 284 data sources. Through the application of standardized criteria, the EPA and stakeholders narrowed the list to 26,000 potential contaminants contained in the 39 most complete and reliable data sources. Further screening caused the EPA to settle on a broad universe of 7,500 contaminants. The EPA then applied health effects and occurrence screening criteria to create a PCCL of roughly 560 contaminants. The final step involved narrowing the PCCL to a draft CCL of 104 contaminants by utilizing an algorithm that scored the chemicals based on four variables designed to isolate the highest public health risks (Drinking Water Contaminant Candidate List 3, 2008).

Following the public release of the preliminary CCL 3, the EPA asked the SAB to review the new protocols. The SAB provided generally positive feedback, calling the new CCL creation process “a more data-driven, systematic approach” (EPA, 2009, p. 3). The SAB, however, made some fairly minor suggestions intended to improve the process. In particular, the

SAB report indicated that the number of contaminants on the draft final CCL was too large to adequately communicate which ones posed the greatest risks. The SAB further recommended that the EPA provide a better explanation of the process by which the agency narrowed the PCCL down to the draft final CCL. The final version of the CCL ended up being slightly larger than the draft version (116 contaminants) and the EPA provided a more elaborate discussion of its decision-making processes (Drinking Water Contaminant Candidate List 3, 2009).

The EPA continued to use the CCL 3 process when it began work on the fourth CCL in 2012 (Request for Nominations, 2012; Drinking Water Contaminant Candidate List 4, 2015). The EPA decided to retain all CCL 3 contaminants, and identify a "universe" of additional contaminants by seeking nominations from the public. Public nominations yielded a list of 59 contaminants that was subsequently narrowed to a universe of 43 following the examination of health effects and occurrence data. Further screening led the EPA to place 20 contaminants on the PCCL. The draft CCL 4, released in early 2015, contained a total of 112 contaminants, most all of which were carried over from the CCL 3 (Drinking Water Contaminant Candidate List 4, 2015).

REGULATORY DETERMINATIONS, 2002-2014

As was previously discussed, the SDWA Amendments required the EPA in five-year intervals to complete regulatory determinations for at least five contaminants listed on the most recent CCL. Doing so, however, first required the EPA to create protocols for selecting the contaminants that posed the greatest health risks and determining whether or not they needed to be regulated. To this end, the EPA enlisted the help of the NRC and NDWAC in developing a scientifically sound regulatory determinations process. The NRC (1999) suggested a particularly elaborate three-step process for making the required regulatory determinations that involved 1) a preliminary examination of each CCL contaminant in which existing data on health effects, exposure, and treatment options is reviewed, 2) a more detailed risk assessment for each contaminant, and 3) a separate decision document for each contaminant detailing whether or not it should be considered for regulation.

The NDWAC protocol also called for a three-step process more closely related to the three enumerated statutory criteria discussed in the preceding section. To assess adverse health effects, the NDWAC recommended that the EPA use scientific risk assessments to create a numerical characterization of overall health risk for each CCL contaminant known as the Health Reference Level (or HRL). During the second step of the protocol, the EPA would then use occurrence data to determine the percentage of US public water systems reporting detections of contaminants at levels above half of the HRL. In the final step, the EPA would determine whether regulating the contaminant presented a meaningful opportunity for health risk reduction by estimating the portion of the population exposed to the contaminant at levels above half of the HRL (Announcement of Preliminary Regulatory Determinations, 2002).

The process ultimately chosen by the EPA incorporated elements of both protocols, but was applied to a fairly limited set of contaminants (Announcement of Preliminary Regulatory Determinations, 2002). Rather than evaluating all 60 contaminants contained on the first CCL, the EPA chose to only work with the list of 20 regulatory priority contaminants. The EPA then narrowed the list down to nine contaminants after determining that inadequate monitoring methods and health and occurrence data existed for the others on the list. The EPA evaluated the nine contaminants using the three-step process recommended by the NDWAC, ultimately deciding not to regulate any of them at that time (Announcement of Regulatory Determinations, 2003). Following the completion of the second CCL, the EPA used a process for selecting and evaluating contaminants for regulatory determinations that roughly mirrored the relatively limited process used after the first CCL (Regulatory Determinations, 2007, 2008).

The limited scope of the first two regulatory determinations ultimately drew criticism from a range of external actors. In particular, a GAO (2011) report criticized the EPA for choosing contaminants for regulatory determination based on the availability of data rather than on a systematic assessment of the appropriate risk priorities. The GAO further criticized the agency for failing to fully explain its decision-making processes and disclose the limitations of some it its analytical techniques. When the US Senate followed up on the report with a hearing later that year, a high-ranking EPA official conceded the inadequacies of its current regulatory determinations process, and reassured legislators that the agency was "continually learning from each iteration of this process and … currently applying lessons learned from previous determinations" (*Oversight Hearing,* 2011, p. 2). The agency also indicated that it would consult with an independent panel of scientists to improve its regulatory determinations protocols.

For the third round of regulatory determinations, the EPA modified existing protocols to produce a more rigorous and comprehensive three-phase process. During the first phase, the EPA analyzed all 116 contaminants contained on the third CCL to identify a short list of contaminants for which there was sufficient health and occurrence data to make a sound regulatory determination (Announcement of Preliminary Regulatory Determinations, 2014, p. 62721). During phase two, the EPA then formulated draft HRLs for 35 of the original 116 contaminants, and compared them against occurrence data to identify only those contaminants that "occur at levels and frequencies of public health concern" (62722). Of the 35 contaminants considered during the second phase, 5 were passed on to the third phase, where EPA staff derived a final HRL. After further analysis, EPA staff recommended to the Administrator that one of the contaminants be regulated (62727).

CONCLUSION

In the two decades since the passage of the SDWA Amendments, the risk management paradigm has become firmly entrenched in the EPA's implementation protocols. Taking the advice of multiple external advisory groups, the EPA has developed elaborate science-based routines for identifying new drinking water contaminants, and determining which ones pose the most significant threats to human health. Although there was a learning curve along the way, there is little doubt that these routines constitute a good faith attempt to put risk management principles into practice.

There may, however, be some room for doubt as to whether risk management principles adequately protect the public health. For one thing, the SDWA Amendments of 1996 appear to provide the EPA with little incentive to regulate new contaminants. This is the case because the law only requires that the EPA consider five new contaminants for regulation every five years, and places the burden of proof on the agency to demonstrate that regulation of a contaminant will actually yield substantial public health benefits. The end result is that the agency's attention is drawn to the relatively small handful of contaminants that have been the subject of extensive public health research. Certainly, this approach serves the short-term interests of state and local governments seeking to avoid the compliance costs associated with regulating new contaminants. Nevertheless, it may cause the EPA to overlook potentially dangerous contaminants because the available research remains inconclusive.

In spite of these limitations, the SDWA, as amended in 1996, generally remains in good standing with EPA officials, state water providers, and members of Congress on both sides of the political spectrum. None of these parties have publicly demonstrated any interest in significantly altering the risk management paradigm, or the federal-state relationship that has served as the foundation of US drinking water regulation for more than a century.

Chapter 6

CONCLUSION

The narrative presented in the preceding chapters suggests that the history of US drinking water regulation is a complex tale. Government officials, existing policies and institutions, discourses, and interest groups of various kinds were among the forces that influenced the development of drinking water policy between 1914 and 2015. Which among these variables had the most significant impact on the content and direction of policy? To answer that question, it is necessary to return to the theoretical perspectives introduced at the beginning of this study.

THE SOCIETY-CENTERED PERSPECTIVE

The society-centered perspective predicts that policy change is most directly driven by societal demand. While societal demand may certainly be indicated by shifts in public opinion polls, this alone does not determine the direction of public policy in scientifically complex issue areas. Rather, the societal perspective is vindicated when organized political interests succeed in determining not only which problems remain on the public agenda, but also the solutions that are ultimately enacted. Furthermore, environmental policy scholars who apply some variation of the society-centered perspective tend to focus on two broad categories of groups:

public interest groups (particularly environmental groups) and economic interests that are directly affected by environmental policies. Public interest groups and economic interests are generally portrayed as being in opposition to one another (Rosenbaum, 2014), though some studies indicate that periods of compromise and interest convergence do occur (Bosso, 1988; Gonzalez, 2001).

Admittedly, the public interest/economic interest dichotomy is difficult to apply to the case of drinking water regulation. The "economic interests" directly affected by drinking water policy are mostly either publicly-owned or private non-profit water systems that claim to be in the business of protecting the public health. Furthermore, water providers have a long history of working cooperatively with state and federal regulators to achieve this goal. Nevertheless, there is also good reason to believe that water providers can be viewed as a set of economic interests with similar incentives to those held by the profit-seeking industries (energy and chemical companies, for instance) that typically oppose environmental regulations. Like other industries, water providers have organized themselves into trade associations and succeeded in gaining a strong voice in the drinking water policy process. In addition, water providers share many of the same concerns about regulation as other industries—that excessive compliance costs will undermine the viability of their enterprise. As the findings of this study suggest, water industry groups vigorously opposed costly federal regulations throughout much of the 1980s and 1990s. Therefore, for the remainder of this discussion, it is assumed that the traditional public interest/economic interest dichotomy is applicable to the context of drinking water regulation.

The data presented in this study suggests that societal interests had at least a moderate impact on US drinking water policy between 1914 and 2015. When the PHS was in charge of setting drinking water policy prior to 1970, water industry interests were directly involved in policy formulation. The PHS standard setting committees brought together researchers and water professionals from around the country to review scientific evidence and formulate quantitative standards for public health protection. This approach was probably inevitable for at least two reasons.

First, although the PHS possessed respected research capacities, drinking water issues were not its main focus by any means. The agency did not have an office solely dedicated to drinking water issues, nor was its staff engaged in extensive research on drinking water issues on an ongoing basis. Second, the PHS did not possess the statutory authority to mandate a binding set of drinking water standards for the entire nation. Thus, achieving consensus was essential if the PHS wanted to produce drinking water standards that were broadly influential.

After 1970, the influence of societal groups on drinking water quality was variable and somewhat hard to measure. Certainly, drinking water issues surfaced in Congress in the early 1970s—a time period in which a wave of new environmental groups was raising public consciousness about a host of environmental issues. At the same time, however, there is little evidence that these national environmental groups were particularly focused on drinking water issues. Table 1 provides a tabulation of the types of groups that participated in congressional hearings dealing with drinking water legislation between 1971 and 1995. During the years when Congress was considering passage of the original SDWA (1971-1974), environmental advocacy groups were barely a part of the conversation. Although the EDF did conduct a high profile study of Mississippi River water contamination in Louisiana, this study was released after drinking water legislation had been under development in Congress for years. Furthermore, while the study may certainly have eased passage of the final legislation, it did not directly influence the content of the legislation. Therefore, the data provides an insufficient basis for concluding that citizen groups were influential in the development of drinking water legislation in the 1970s.

The evidence of environmental group influence in the 1980s and 1990s is no more conclusive. Table 1 suggests that environmental groups and other types of citizen groups became more involved in legislative debates in the 1980s. Furthermore, the SDWA Amendments of 1986 likely reflected the preferences of many environmental groups for a statute that held the EPA and state governments to stricter implementation requirements. In the 1990s, however, when these new requirements proved impractical and expensive to implement, Congress moved in a different direction with seemingly little regard for the preferences of environmental groups.

Evidence supporting industry involvement in the formulation of drinking water policy between 1971 and 1996 is equally mixed. Though Table 1 suggests that representatives of the water industry were vocal participants in legislative debates throughout this period, the evidence from the historical narrative suggests that their influence varied over time. In the early 1970s, representatives of the drinking water industry cautiously supported national standards, but why wouldn't they? The bill that ultimately advanced through Congress largely left in tact the existing drinking water policy regime developed through PHS and industry cooperation during the preceding sixty years. While some industry trade associations expressed concerns about the potential costs associated with mandatory national standards, the final legislation gave the EPA plenty of opportunity to take costs into consideration when formulating regulations. Therefore, while it is not clear that industry groups directly shaped the content of the SDWA, Congress certainly managed to produce a law that was palatable to the industry as a whole.

Table 1. Witnesses Testifying in Drinking Water Hearings, 1971-1995

	1971-1973	1982-1985	1991-1995	Total
EPA	11	14	9	34
Other Fed. Agency	7	4	4	15
Congress	9	7	3	19
State Gov.	1	35	26	62
Water Industry	16	26	18	60
Water/Pub. Health Professionals	12	4	1	17
Environmental Groups	0	15	8	23
Other Citizen Groups	3	7	0	10
Non-Water Industries	3	11	0	14
Researchers	6	6	1	13

Source: Congressional Information Service (CIS) hearing abstracts

During the 1980s, by contrast, industry influence seemed to wane a little. While industry concerns about the costs associated with EPA regulations brought drinking water issues back onto the agenda in the early 1980s, industry groups had little success shaping the content of the SDWA Amendments of 1986. In the 1990s, Congress moved in a direction that was favorable to the industry, though it is not clear that industry had any more luck influencing the content of legislation than they had in the past. It seems more likely that both the industry and members of Congress signed on to the risk management proposals advanced by the EPA, which heavily shaped the contents of the SDWA Amendments of 1996.

THE STATE-CENTERED PERSPECTIVE

The state-centered perspective asserts that the direction of public policy is mainly determined by existing institutional arrangements and the autonomous goals and policy preferences held by state officials. Even when policy development ensues in response to public pressure and societal crises, state officials frequently take the lead in determining the contents of public policy. Furthermore, policy formulation decisions, whether or not made directly by state officials, are frequently shaped by policy legacies that make fundamental departures from the status quo seem impractical.

The evidence presented in preceding chapters provides strong evidence that state actors and institutions were highly influential throughout most of the period under study. As early as the 1860s, state governments began actively investigating drinking water filtration techniques, if only to improve the appearance and taste of drinking water. When typhoid fever outbreaks led to public health crises in many urban areas, state boards of health were prepared to utilize existing scientific knowledge to fashion treatment and disinfection technologies that made water potable once again. By the second decade of the twentieth century, drinking water treatment techniques were well established and becoming widely utilized.

In 1914, the federal government entered the fray when it approved the first national quality standard for drinking water served in interstate commerce. Although the federal government was not in the business of promulgating a broad array of national regulatory standards at that time, it was willing to formulate the drinking water quality standard for a number of reasons. First, the regulation of drinking water quality seemed to fit with the traditional "quarantine" power that Congress believed it possessed as part of its power to regulate interstate commerce. Second, by 1914, the federal government had already established a public health bureaucracy that could serve as an institutional home for drinking water issues. Taken together, these institutional realities made it possible for the PHS to initiate and direct the development of the first national drinking water standard.

Although the 1914 drinking water standard was only legally binding on water served in interstate commerce, the water industry informally adopted it as an industry-wide quality standard. This unintended but happy consequence caused the PHS to revise and expand its drinking water standards several times over the next four decades. Along the way, PHS scientists played a key role, particularly with respect to the development of quantitative standards for chemical contaminants. Undoubtedly, PHS officials were able to act with a degree of autonomy because the standards they created remained largely voluntary for most water providers. Nevertheless, what evolved over time was an informal regulatory regime for drinking water protection in which the PHS set national standards, while the states retained enforcement responsibilities. When Congress took up the issue of drinking water regulation in the 1970s, this policy legacy served as an important set of building blocks.

Furthermore, state actors and institutional arrangements played key roles in the development of legislation between 1971 and 1996. In the 1970s, drinking water issues found their way onto the national agenda not because of societal pressure, but because of a PHS report that identified inconsistencies in drinking water quality across states and localities. These findings motivated Reps. Rogers and Robison to pursue national drinking water legislation in spite of the early reluctance expressed by both the EPA and the water industry. Eventually, the EPA was able to forge a consensus by developing a legislative proposal that institutionalized key components of the existing drinking water policy regime. The EPA's consensus proposal certainly reflected the interests of the drinking water industry. It must be kept in mind, however, that preexisting policy arrangements heavily shaped industry interests. Therefore, it seems like a reasonable interpretation of the evidence to conclude that state actors and institutions played a more significant role in the development of the SDWA of 1974 than interest groups.

The same can be said about the amendments to the SDWA passed in both 1986 and 1996. During the early 1980s, industry concerns about the costs associated with EPA regulations helped bring drinking water issues back onto the national agenda. However, industry concerns about costs mainly received a hearing because Senator Gorton shared these concerns. When Senator Durenberger assumed the chairmanship of the Subcommittee on Toxic Substances and Environmental Oversight, however, the political landscape changed. Senator Durenberger's personal desire to prioritize public health protection over economic costs was reflected in the stringent requirements placed on the EPA under the SDWA Amendments of 1986.

During the 1990s, drinking water issues emerged on the agenda once again as a result of a confluence of events that included renewed industry concerns about costs combined with public concern centered on the drinking water crisis in Milwaukee. The contents of the SDWA Amendments of 1996, however, were most heavily shaped by proposals developed by the EPA. The EPA developed a series of proposals rooted in the "risk management" approach to pollution control that revolved around three core principles: pollution prevention, risk-based priority setting, and public involvement and education. These principles, articulated in several EPA policy documents, served as the basis for most of the major provisions contained in the final legislation.

The Policy Discourse Perspective

Scholars of policy discourse assert that ideas and language are capable of altering the way political actors view societal problems and assess potential solutions. A particular discourse becomes dominant when actors within a policy community must utilize the language and practices associated with the discourse to retain credibility. Truly dominant discourses also become institutionalized within the state through legal enactments and governmental routines.

When the history of drinking water regulation is viewed through these lenses, it becomes evident that discourse played a role in shaping policy outcomes at certain times during the period under study. Although there was no explicit drinking water policy discourse prior to 1970, it does appear that PHS staff and members of the various drinking water standard-setting committees adhered to a common set of implicit assumptions. Among these assumptions was the belief that public health protection was the singular goal of federal drinking water policy. Furthermore, it was assumed that this goal could be achieved by using the best available scientific evidence to establish "safe" levels of human exposure to known drinking water contaminants. Economic costs were not a consideration for federal regulators because the standards they promulgated remained largely voluntary. As long as water providers and state regulators retained the ability to meet federal standards on their own terms, they were in control of the costs associated with compliance. This reality gave federal regulators the freedom to set standards that were protective of public health while avoiding conflict with the water industry over compliance costs or the proper way to interpret scientific evidence.

Passage of the SDWA in 1974 upended these accepted beliefs and sent the drinking water policy community on a two-decade search for a new guiding philosophy. Once national standards became mandatory, it was no longer politically feasible for federal regulators to formulate highly protective regulations without reference to compliance costs. In the early 1980s, however, few within the drinking water community could figure out how to balance economic costs and public health protection in a way that produced consensus. The controversy surrounding GAC filtration technology brought this reality into sharp relief. The GAC mandate was the first major time federal regulators imposed a costly treatment technology on public water systems. Although EPA officials never conducted a formal cost-benefit analysis, they seemed confident that the potential public health benefits outweighed the costs. Nevertheless, the political pushback they received from the water industry was unprecedented up to that time. When Ann Buford became EPA Administrator, her staff chose to reverse the GAC mandate based primarily on cost considerations, which in turn led to a backlash from members of Congress who were concerned about the public health implications of this action. Ultimately, this backlash produced the stringent regulatory mandate contained in the SDWA Amendments of 1986, which prioritized public health protection over costs.

William Ruckleshaus, Buford's successor as EPA Administrator, promoted the risk management paradigm as a middle way that would allow members of the environmental policy community to reconcile public health protection with economic costs. The key, for Ruckleshaus, was to inject science-based priority setting into the regulatory process. Public health protection would remain the predominant value, but the agency would use quantitative risk assessment to help focus its limited resources on the most serious public health threats. Furthermore, Ruckleshaus believed that when stakeholders became better educated about risk tradeoffs, political consensus would be more likely to develop. Industries would come to accept some compliance costs because costs were held in check and regulatory decisions were based in sound science. Environmental and public health activists would learn to accept marginal risks once they understood that the most serious risks were being identified and dealt with.

Over the next several years, Ruckleshaus' risk management vision took on the characteristics of a policy discourse. Risk management principles were elaborated by EPA officials in subsequent reports and, in the early 1990s, gained currency among both Republicans and Democrats in Congress who were actively debating SDWA reform proposals. The legislation that was ultimately passed in 1996 gave the EPA clear instructions to evaluate and regulate new drinking water contaminants according to risk management principles. Over the following two decades, EPA officials worked closely with multiple advisory groups to create and put into practice elaborate risk-based protocols for identifying and evaluating contaminants that posed potential public health threats. As a result, risk management principles and concepts remain deeply embedded in the practice of federal drinking water regulation to this day.

At the same time, however, the case of drinking water regulation demonstrates the limits of policy discourse as a force for policy change. Although the risk management discourse changed the conversation about drinking water regulation, the resulting policy changes were largely incremental in nature. The risk management provisions contained in the SDWA Amendments of 1996 only impacted new contaminants; the MCLs promulgated prior to that time largely remain in place without amendment. Given that the EPA has only chosen to regulate one new contaminant since 1996, it is difficult to conclude that risk management principles have had a revolutionary impact on the direction of US drinking water policy.

FINAL THOUGHTS

After carefully reviewing the relative explanatory power of each theoretical perspective, it seems evident that the state-centered approach best explains changes in drinking water policy over the last century. Between 1914 and 1970, drinking water policy was heavily shaped by the institutional realities of the US federal system. Federal policy innovations after 1970 largely built upon, rather than supplanted, the policy legacies established prior to that time. Furthermore, the contents of federal legislation, particularly in the 1980s and 1990s, largely reflected ideas and preferences advanced by state officials rather than societal groups. Without question, the risk management discourse played an important role in building support for the SDWA Amendments of 1996. Nevertheless, it must be kept in mind that risk management ideas were developed and advanced by state officials who sought to reframe the conversation over

US pollution control laws. For this reason, it is hard to view discourse as an independent variable within the context of US drinking water regulation.

Do these findings have broader implications for the study of US environmental and public health policy? Perhaps not. Few other studies of US environmental policy have found the state-centered perspective to hold a high degree of explanatory power. It may very well be the case that the politics surrounding drinking water policy are different than those found in other environmental policy domains. At the same time, however, the findings contained in this study should caution policy scholars against treating US environmental policy as a single, unified policy domain that can adequately be explained by a single theory. As Klyza (1992, p. 176) correctly noted, "our task should not be to crown a non-contingent theoretical ruler, but rather to determine which characteristics make a particular theoretical approach best suited to wear the crown of explanation for a particular case."

REFERENCES

Altman, L. K. (1993, April 20). Outbreak of disease in Milwaukee undercuts confidence in water. *The New York Times*, p. C3.

Andrews, R. N. L. (1999). Managing the environment, managing ourselves: A history of American environmental policy. New Haven: Yale University Press.

Announcement of the Draft Drinking Water Contaminant Candidate List, 62 Fed. Reg. 52104 (1997).

Announcement of the Draft Drinking Water Contaminant Candidate List, 63 Fed. Reg. 10274 (1998).

Announcement of Preliminary Regulatory Determinations for the Contaminants on the Third Drinking Water Contaminant Candidate List, 79 Fed. Reg. 62716 (2014).

Announcement of Preliminary Regulatory Determinations for the Priority Contaminants on the Drinking Water Contaminant Candidate List, 67 Fed. Reg. 38222 (2002).

Announcement of Regulatory Determinations for the Priority Contaminants on the Drinking Water Contaminant Candidate List, 68 Fed. Reg. 42898 (2003).

Anton, T. J. (1989). American federalism and public policy: How the system works. New York: Random House.

Baker, M. N. (1948). The quest for pure water: The history of water purification from the earliest records to the twentieth century. New York: American Water Works Association.

Baumgartner, F. R. & Jones, B. D. (2009). Agendas and instability in American politics (2nd ed.). Chicago: University of Chicago Press.

Birkland, T. A. (1997). After disaster: Agenda setting, public policy, and focusing events. Washington, D.C.: Georgetown University Press.

Bosso, C. J. (1987). Pesticides and politics: The life cycle of a public issue. Pittsburgh: University of Pittsburgh Press.

Bosso, C. J. (1988). Transforming adversaries into collaborators: Interest groups and the regulation of chemical pesticides. *Policy Sciences*, *21*, 3-22.

Bosso, C. J. (2005). Environment, Inc.: From grassroots to beltway. Lawrence: University Press of Kansas.

Brown, A. E. (1987). The politics of airline deregulation. Knoxville: University of Tennessee Press.

Bryner, G. C. (1987). Bureaucratic discretion: Law and policy in federal regulatory agencies. New York: Pergamon Press.

Bryner, G. C. (1995). Blue skies, green politics: The Clean Air Act of 1990 and its implementation (2nd ed.). Washington, DC: Congressional Quarterly Press.

Cahn, M. A. (1995). Environmental deceptions: The tension between liberalism and environmental policymaking in the United States. Stony Brook: State University of New York Press.

Carpenter, D. P. (2001). The forging of bureaucratic autonomy: Reputations, networks, and policy innovation in executive agencies, 1862-1928. Princeton: Princeton University Press.

Control of Organic Chemical Contaminants in Drinking Water, 43 Fed. Reg. 5756 (1978).

Cook, B. J. (1988). Bureaucratic politics and regulatory reform: The EPA and emissions trading. New York: Praeger.

Cosgrove, J. J. (1909). *History of sanitation.* Pittsburgh: Standard Sanitary Manufacturing Company.

Derthick, M. & Quirk, P. J. (1985). The politics of deregulation. Washington, DC: Brookings Institution.

Drinking Water Contaminant Candidate List 2 – Notice, 69 Fed. Reg. 17406 (2004).

Drinking Water Contaminant Candidate List 2 – Final Notice, 70 Fed. Reg. 9071 (2005).

Drinking Water Contaminant Candidate List 3 – Draft, 73 Fed. Reg. 9628 (2008).

Drinking Water Contaminant Candidate List 3 – Final, 74 Fed. Reg. 51850 (2009).

Drinking Water Contaminant Candidate List 4 – Draft, 80 Fed. Reg. 6076 (2015).

Dryzek, J. S. (2005). The politics of the earth: Environmental discourses (2nd ed.). New York: Oxford University Press.

Eisner, M. A. (2000). Regulatory politics in transition (2nd ed.). Baltimore: Johns Hopkins University Press.

Ellms, J. W. (1917). Water purification. New York: McGraw-Hill.

Federal Safe Drinking Water Act – Oversight: Hearing before the Subcommittee on Health and the Environment of the Committee on Interstate and Foreign Commerce, House of Representatives, 95th Cong. 1 (1978).

Fiorino, D. J. (1990). Can problems shape priorities?: The case of risk-based environmental planning. *Public Administration Review*, *50*, 82-90.

Fiorino, D. J. (2001). Environmental policy as learning: A new view of an old landscape. *Public Administration Review*, *61*, 322-334.

Flint Water Advisory Task Force. (2016). *Final report.* Retrieved from: https://www.michigan.gov/documents/snyder/FWATF_FINAL_REPORT_21March2016_517805_7.pdf.

Gonzalez, G. A. (2001). Corporate power and the environment: The political economy of US environmental policy. Washington, DC: Rowman and Littlefield.

Gormley, W. T. (1986). Regulatory issue networks in a federal system. *Polity*, *18*, 595-620.

Graham, D. A. (2016, January). What did the governor know about Flint's water, and when did he know it? The Atlantic. Retrieved from: https://www.theatlantic.com/politics/archive/2016/01/what-did-the-governor-know-about-flints-water-and-when-did-he-know-it/423342/.

Green, M. J. (Ed.) (1973). The monopoly makers: Ralph Nader's study group report on regulation and competition. New York: Grossman.

Habicht II, F. H. (1994). EPA's vision for setting national environmental priorities. In A.M. Finkel & D. Golding (Eds.), Worst things first: The debate over risk-based national environmental priorities (pp. 33-46). Washington, DC: Resources for the Future.

Hajer, M. (1995). The politics of environmental discourse: Ecological modernization and the policy process. Oxford: Oxford University Press.

Harden, V. A. (1986). Inventing the NIH: Federal biomedical research policy, 1887-1937. Baltimore: Johns Hopkins University Press.

Harris, R. A. & Milkis, S. M. (1996). The politics of regulatory change: A tale of two agencies (2nd ed.). New York: Oxford University Press.

Hazen, A. (1900). The filtration of public water supplies (3rd ed.). New York: Wiley and Sons.

Hazen, A. (1914). Clean water and how to get it (2nd ed.). New York: Wiley and Sons.

Hoornbeek, J. A. (2011). Water pollution policies and the American states: Runaway bureaucracies or congressional control? Albany, NY: State University of New York Press.

Hopkins, O. C. (1961). Attitude of US Public Health Service in development of drinking water standards, 1961. *Journal of the American Water Works Association*, *53*, 946-950.

Horwitz, R. B. (1989). The irony of regulatory reform: The deregulation of American telecommunications. New York: Oxford University Press.

Jasanoff, S. (1990). The fifth branch: Science advisors as policymakers. Cambridge, MA: Harvard University Press.

Jennings, C. A. (1923). The disinfection of public water supplies and its relation to public health. *Journal of the American Water Works Association*, *10*, 127-138.

Jochim, A. E. & May, P. J. (2010). Beyond subsystems: Policy regimes and governance. *Policy Studies Journal*, *38*, 303-327.

Johnson, G. A. (1914). Present day water filtration practice. *Journal of the American Water Works Association*, *1*, 31-80.

Jones, C. O. (1974). Speculative augmentation in federal air pollution policymaking. *Journal of Politics*, *36*, 438-464.

Kamieniecki, S. (2006). Corporate America and environmental policy: How often does business get its way? Palo Alto: Stanford University Press.

Ketcham-Colwill, J. (1986). Safe drinking water law toughened. *Environment*, *28*(7), 5, 42-43.

Kimm, V. J., Kuzmack, A. M. & Schnare. D. W. (1981). Waterborne carcinogens: A regulator's view. In R.W. Crandall & L.B. Lave (Eds.), The scientific basis of health and safety regulation (pp. 229-249). Washington, DC: Brookings Institution.

Kirkwood, J. P. (1869). Report on the filtration of river waters for the supply of cities as practiced in Europe. New York: Van Nostrand.

Klyza, C. M. (1992). A window of autonomy: State autonomy and the Forest Service in the early 1990s. *Polity*, *25*, 173-196.

Klyza, C. M. (1996). Who controls public lands?: Mining, forestry, and grazing policies, 1870-1990. Chapel Hill: University of North Carolina Press.

Klyza, C. M. & Sousa, D. (2008). American environmental policy, 1990-2006: Beyond gridlock. Cambridge, MA: MIT Press.

Kotin, P. & Hueper, W. C. (1955). Relationship of industrial carcinogens to cancer in the general population. *Public Health Reports*, *70*, 331-333.

Kraft, M. E. & Kamieniecki, S. (Eds.). (2007). Business and environmental policy: Corporate interests in the American political system. Cambridge, MA: MIT Press.

Lead Contamination Control Act, P.L. 100-572, October 31st, 1988.

Lead contamination: Hearings before the Subcommittee on Health and Environment of the Committee on Energy and Commerce, House, 100th Cong. 1 (1988).

Lijphart, A. (1971). Comparative politics and the comparative method. *American Political Science Review*, *65*, 682-693.

Lin, J. C. F., Rutter, J. & Park, H. (2016, January 21). Events that led to Flint's water crisis. *New York Times.* Retrieved from http://www.nytimes.com/interactive/2016/01/21/us/flint-lead-water-timeline.html

Litfin, K. (1994). Ozone discourses: Science and politics in global environmental cooperation. New York: Columbia University Press.

Marx, J. (1974). Drinking water: Another source of cancer? *Science*, *186*(4166), 809-811.

Matukaitis, J. M. (1997). The emerging recognition of Cryptosporidium as a health hazard. *Journal of Community Health Nursing*, *14*(3), 135-140.

McCraw, T. K. (1985). Prophets of regulation. Cambridge, MA: Belknap Press.

McDermott, J. H. (1973). Federal drinking water standards: Past, present, and future. *Journal of the Environmental Engineering Division—ASCE*, EE4, 469-478.

McLaughlin, A. J. (1914). What is safe drinking water? *Public Health Reports*, *26*, 1686-1694.

Merrill, R. A. (1997). Food safety regulation: Reforming the Delaney clause. *Annual Review of Public Health*, *18*, 313-340.

Milazzo, P. C. (2006). Unlikely environmentalists: Congress and the Clean Water Act, 1945-1972. Lawrence, KS: University Press of Kansas.

Moe, T. M. (1989). The politics of bureaucratic structure. In J.E. Chubb & P.E. Peterson (Eds.), Can the government govern? (pp. 267-329). Washington, D.C.: Brookings Institution.

National Drinking Water Advisory Council. (2004). Report on the CCL classification process to the US Environmental Protection Agency. Washington, D.C.: National Drinking Water Advisory Council.

National Interim Primary Drinking Water Regulations: Trihalomethanes, 48 Fed. Reg. 8406 (1983).

National Interim Primary Drinking Water Standards, 40 Fed. Reg. 59566 (1975).

National Research Council. (1977). Drinking water and health (vol. 1). Washington, DC: National Academies Press.

National Research Council. (1999). Setting priorities for drinking water contaminants. Washington, D.C.: National Academies Press.

National Research Council. (2001). Classifying drinking water contaminants for regulatory consideration. Washington, D.C.: National Academies Press.

Olson, M. (1965). The logic of collective action. Cambridge, MA: Harvard University Press.

Organic Chemical Contaminants: Control Options in Drinking Water, 41 Fed. Reg, 28991 (1976).

Oversight Hearing on the Environmental Protection Agency's implementation of the Safe Drinking Water Act's Unregulated Drinking Water Contaminant Program: Hearing before the Committee on Environment and Public Works, Senate, 112th Cong. (2011). (Testimony of Robert Perciasepe). Retrieved at http://www.epw.senate.gov/public/index.cfm/2011/7/full-committee-hearing-entitled-oversight-hearing-on-the-environmental-protection-agencys-implementation-of-the-safe-drinking-water-acts-unregulated-drinking-water-contaminants-program.

Page, T., Harris, R. H. & Epstein, S. S. (1976). Drinking water and cancer mortality in Louisiana. *Science*, *193*(4247), 55-57.

Pontius, F. W. (1993). Protecting the public against Cryptosporidium. *Journal of the American Water Works Association*, *85*(8), 18, 22, 122-123.

Pontius, F. W. (2003). History of the Safe Drinking Water Act. In F.W. Pontius (Ed.) *Drinking water regulation and health* (pp. 71-103). New York: Wiley-Interscience.

Potable waters: Hearing before the Committee on Commerce, Senate, 92nd Cong. 1 (1972).

Regulatory Determinations Regarding Contaminants on the Second Drinking Water Contaminant Candidate List – Preliminary Determinations, 72 Fed. Reg. 24016 (2007).

Regulatory Determinations Regarding Contaminants on the Second Drinking Water Contaminant Candidate List, 73 Fed. Reg. 44251 (2008).

Request for Nominations of Drinking Water Contaminants for the Fourth Contaminant Candidate List, 77 Fed. Reg. 27057 (2012).

Rosenbaum, W.A. (2014). Environmental politics and policy (9th ed.). Washington, DC: Congressional Quarterly Press.

Ruckleshaus, W.D. (1985). Risk, science, and democracy. *Issues in Science and Technology*, *1* (3), 19-38.

Russell, C.S. (Ed.). (1978). Safe drinking water: Current and future problems. Washington, DC: Resources for the Future.

Safe drinking water: Hearing before the Subcommittee on Public Health and Environment of the Committee on Interstate and Foreign Commerce, House of Representatives, 92nd Cong. 1 (1971).

Safe drinking water: Hearing before the Subcommittee on Toxic Substances and Environmental Oversight of the Committee on Environment and Public Works, Senate, 97th Cong. 1 (1982).

Safe Drinking Water Act of 1973: Hearing before the Subcommittee on Environment of the Committee on Commerce, Senate, 93rd Cong. 1 (1973).

Safe Drinking Water Act of 1974, P.L. 93-523, December 16th, 1974.

Safe Drinking Water Act Amendments: Hearing before the Subcommittee on Health and the Environment of the Committee on Interstate and Foreign Commerce, House, 95th Cong. 1 (1977).

Safe Drinking Water Act Amendments of 1986, P.L. 99-339, June 19th, 1986.

Safe Drinking Water Act Amendments of 1993: Hearing before the Committee on Environment and Public Works, Senate, 103rd Cong. 1 (1993).

Safe Drinking Water Act Amendments of 1996, P.L. 104-182, August 6th, 1996.

Safe Drinking Water Act oversight: Hearing before the Subcommittee on Environmental Pollution of the Committee on Environment and Public Works, Senate, 95th Cong. 1 (1978).

Safe Drinking Water Amendments of 1985: Hearing before the Subcommittee on Toxic Substances and Environmental Oversight of the Committee on Environment and Public Works, Senate, 99th Cong. 1 (1985).

Sedgwick, W. T. (1902). Principles of sanitary science and the public health. New York: Macmillan.

Sellers, C. (1997). Discovering environmental cancer: Wilhelm Hueper, post-World War II Epidemiology, and the vanishing clinician's eye. *American Journal of Public Health*, *87*, 1824-1835.

Skocpol, T. (1985). Bringing the state back in: Strategies of analysis in current research. In P.B. Evans, D. Rueschemeyer, & T. Skocpol (Eds.), Bringing the state back in (pp. 3-43). Cambridge, UK: Cambridge University Press.

Skowronek, S. (1982). Building a new American state: The expansion of national administrative capacities, 1877-1920. Cambridge, UK: Cambridge University Press.

Stockington, H. E. & Woodward, R. L. (1958). Toxicologic methods for establishing drinking water standards. *Journal of the American Water Works Association*, *50*, 515-529.

Swift, E. K. & Brady, D. W. (1994). Common ground: History and theories of American politics. In L.C. Dodd & C. Jillson (Eds.), The dynamics of American politics (pp. 83-104). Boulder, CO: Westview Press.

Tarlock, A. D. (1997). Safe drinking water: A federalism perspective. *William and Mary Environmental Law and Policy Review*, *21*, 233-263.

Tarr, J., Curley, J., McMichael, F. C. & Yosie, T. (1984). Water and wastes: A retrospective assessment of wastewater technology in the United States, 1800-1932. *Technology and Culture*, *25*, 226-263.

U.S. Environmental Protection Agency. (1972). Industrial pollution of the lower Mississippi River in Louisiana. Dallas: Region VI Surveillance and Analysis Division.

U.S. Environmental Protection Agency. (1986). Reducing lead in drinking water: A benefit analysis. Washington, DC: Office of Policy, Planning, and Evaluation.

U.S. Environmental Protection Agency. (1987). Unfinished business: A comparative assessment of environmental problems. Washington, DC: Office of Policy, Planning, and Evaluation.

U.S. Environmental Protection Agency. (1990). Reducing risk: Setting priorities and strategies for environmental protection. Washington, DC: Science Advisory Board, Relative Risk Reduction Strategies Committee.

U.S. Environmental Protection Agency. (1993). Administration recommendations for Safe Drinking Water Act reauthorization. Washington, DC: Office of Water.

U.S. Environmental Protection Agency. (1995). Strengthening the safety of our drinking water: A report on progress and challenges and an agenda for action. Washington, DC: Office of Water.

U.S. Environmental Protection Agency. (1996). The conceptual approach for contaminant identification. Washington, D.C.: Office of Water.

U.S. Environmental Protection Agency. (2009). SAB advisory on EPA's draft third contaminant candidate list. Washington, D.C.: Science Advisory Board, Drinking Water Committee.

U.S. General Accounting Office. (1973). Improved federal and state programs needed to insure the purity and safety of drinking water in the United States. Washington, DC: U.S. General Accounting Office.

U.S. General Accounting Office. (1982). States' compliance lacking in meeting safe drinking water regulations. Washington, DC: U.S. General Accounting Office.

U.S. General Accounting Office. (1990a). Drinking water: Compliance problems undermine EPA program as new challenges emerge. Washington, DC: U.S. General Accounting Office.

U.S. General Accounting Office. (1990b). Toxic substances: Effectiveness of unreasonable risk standards unclear. Washington, DC: U.S. General Accounting Office.

U.S. General Accounting Office. (1992). Drinking water: Widening gap between needs and available resources threatens vital EPA program. Washington, DC: U.S. General Accounting Office.

U.S. General Accounting Office. (1993). Drinking water: States face increased difficulties in meeting basic requirements. Washington, DC: U.S. General Accounting Office.

U.S. Government Accountability Office. (2011). EPA should improve implementation of requirements on whether to regulate additional contaminants. Washington, D.C.: U.S. General Accounting Office.

U.S. Public Health Service. (1914). Bacteriological standard for drinking water: The standard adopted by the Treasury Department for drinking water supplied to the public by common carriers in interstate commerce. *Public Health Reports*, *29*, 2959-2966.

U.S. Public Health Service. (1922). The U.S. Public Health Service Advisory Committee on Official Water Standards. *Public Health Reports*, *37*, 1466-1469.

U.S. Public Health Service. (1925). Report of the Advisory Committee on Official Water Standards. *Public Health Reports*, *40*, 693-722.

U.S. Public Health Service. (1943a). Public Health Service drinking water standards. *Public Health Reports*, *58*, 69-82.

U.S. Public Health Service. (1943b). Manual of recommended water sanitation practice. *Public Health Reports*, *58*, 83-111.

U.S. Public Health Service. (1946). Public health drinking water standards, 1946. *Public Health Reports, 61,* 371-384.

U.S. Public Health Service. (1961). Drinking water standards, 1961. *Journal of the American Water Works Association*, *53*, 935-945.

U.S. Public Health Service. (1962). *Drinking water standards.* PHS publication #956. Washington, DC: Government Printing Office.

U.S. Public Health Service. (1970). Community water supply study: Analysis of national findings. Washington, DC: Bureau of Water Hygiene.

U.S. Treasury Department. (1903). Quarantine laws and regulations of the United States. Washington, DC: Government Printing Office.

U.S. Treasury Department. (1909). Annual report of the Surgeon General of the Public Health and Marine Hospital Service of the United States. Washington, DC: Government Printing Office.

Vogel, D. (1989). Fluctuating fortunes: The political power of business in America. New York: Basic Books.

Weidenbaum, M. L. (1978). On estimating regulatory costs. *Regulation*, *2*(3), 14-17.

Weir, M. (1992). Politics and jobs: The boundaries of employment policy in the United States. Princeton: Princeton University Press.

Weir, M. & Skocpol, T. (1985). State structures and the possibilities for "Keynesian" responses to the Great Depression in Sweden, Britain, and the United States. In P.B. Evans, D. Rueschesmeyer, & T. Skocpol (Eds), Bringing the state back in (pp. 107-168). Cambridge, UK: Cambridge University Press.

Weiss, C. (2007). Defining precaution. *Environment*, *49*(8), 33-36.

Whipple, G. C. (1908). Typhoid fever: Its causation, transmission, and prevention. New York: Wiley and Sons.

Williams, B. A. & Matheny, A. R. (1995). Democracy, dialogues, and environmental disputes: The contested languages of social regulation. New Haven: Yale University Press.

Williams, R. C. (1951). The United States Public Health Service, 1798-1950. Washington, DC: Commissioned Officers Association of the United States Public Health Service.

Wilson, J. Q. (1974). Political organizations. New York: Basic Books.

Wolman, A. (1960). Concepts of policy in the formulation of so-called standards for health and safety. *Journal of the American Water Works Association*, *52*, 1343-1348.

Zarkin, M. (2003). Social learning and the history of US telecommunications policy, 1900-1996. Lewiston, NY: Edwin Mellen Press.

Zarkin, M. (2015). Unconventional pollution control politics: The reformation of the US Safe Drinking Water Act. *Electronic Green Journal, 1*(38). Retrieved from www.escholarship.org/uc/search?entity =uclalib_egj;volume=1;issue=38.

Zarkin, M. (2017). Policy learning mechanisms and the regulation of US drinking water. *Public Organization Review*, *17*, 275-291.

INDEX

H

I

K

L

M

N

P

R

S

T

U

W